Introduction

Intelligent Business Intermediate Video can be used alongside the *Intelligent Business Intermediate* course or as free-standing material.

The fictional drama tells the story of two companies who have come together to launch a new product. The storyline demonstrates many of the features of modern business that students will meet in their everyday lives, such as chairing meetings, giving presentations, running brainstorming sessions, etc.

The video and this Video Resource Book have been designed to be transparent and easy to use. They are both divided into five parts. Each part of the *Video Resource Book* is designed to be used systematically, page by page. Each part has four sections.

Preview

The exercises in these sections are designed to act as an introduction to the part of the video that the students are about to watch. They include exercises which focus on a variety of specific skills demonstrated in the video as well as vocabulary exercises (*Video vocabulary*) which pre-teach or activate key vocabulary which students are about to encounter.

First view

These sections contain exercises that students can do as they watch the video or immediately after they watch it for the first time. Sometimes the students are given comprehension questions which they should be encouraged to read before they watch. These will help them to gain a general understanding of what they see. Multiple-choice or *true* or *false* questions focus on more detailed comprehension.

Second view

In these exercises, the students are asked to watch different sections of the video again and answer more specific questions about the content and language used. They may need to watch each section more than once to answer the questions.

After viewing

These sections give practice in some of the grammar points exemplified in the video. They are linked to the corresponding grammar focus of the *Intelligent Business Coursebook* and *Skills Book*, but can be done equally well by students who are not using these books.

A video script and answer key are supplied. You may like to give students the video script at the end of their work on each part.

Paridon and Bluenet

This story is about cooperation between two companies, Paridon and Bluenet. These two companies are very different in structure and corporate culture. The success of their project will depend on their ability to work together.

Paridon

Paridon is a small company which was started by entrepreneurs Jamie Flacks and Philippe Henetier, who design computer games. They are the Creative Directors of the company. The third member of the management team is Hilary Morrison, the Managing Director. She is a graduate of the London Business School and handles the day-to-day running of the business. Over the last six years, Paridon has grown quickly from two bedrooms and two people to 40 people, an office in central London, and one in Paris. Their most famous products are the computer games Grave Digger and Cyberbug III, which have won several awards.

Jamie and Philippe want the company to develop beyond the production of computer games software, so they have recently turned their attention to the design of hardware. They have invented a music player which incorporates a games machine and a mobile phone with internet access.

Paridon doesn't have the staff or the experience to market this new device effectively, so they have sought the help of Bluenet, a major marketing company. Timing is crucial as they are aware that a rival company is developing something similar. A large trade fair in Chicago will be the ideal place to launch their product, but they will have to move fast to be ready in time.

Bluenet Global Corporation

Bluenet is a large and very well-respected company with its headquarters in New York and branch offices in 26 major cities around the world. It is a leading marketing company, with clients ranging from car manufacturers to holiday companies. Bluenet has a well-established reputation in the worldwide marketing of music and video related state-of-the-art designer hardware, which made it the obvious choice for Paridon when they were looking for someone to help them market their new product.

Paridon have been discussing the project with Penelope Bates, the president of Bluenet's UK division. Penelope and her chief marketing executive Jet Patel realise that there is great potential in Paridon's new device. They can see that Jamie and Philippe are very talented designers and that Paridon has a great future even though it is quite a small company at present.

Jamie Flacks

Jamie is a talented entrepreneur, very keen to expand the business and prepared to take risks. He is very hardworking and has great belief in the products he invents, with a good understanding of what young people are looking for in technological devices. He is tactful when dealing with business associates and clients and, though slightly nervous when speaking in public, nevertheless good at making effective and persuasive presentations.

Philippe Henetier

Philippe moved to London from Paris six years ago when he and Jamie decided to set up Paridon together. He is a talented inventor, who cares deeply about Paridon's reputation and relationship with its clients. This can make him appear a little short-tempered at times. Some people find him a bit long-winded when it comes to meetings, but he is worth listening to because he makes some very good points.

Hilary Morrison

Hilary joined Paridon as its Managing Director shortly after the company was founded. She is calm and efficient and handles the day-to-day running of the company with charm and skill. Although she has little understanding of the technology involved in developing Paridon's products, she has great respect for what Jamie and Philippe have achieved and a strong commitment to the company. She manages to keep firm control over the practicalities of the business without getting in the way of Jamie and Philippe's creativity.

Chuck Fenno

Chuck is the CEO of Bluenet and is based in their New York headquarters. He travels around the world visiting Bluenet's overseas offices. At the moment he is in London liaising with Penelope Bates over the Paridon project.

Chuck is a forceful leader who knows what he wants and how to get it. Some people find him rather direct, and he can be impatient if he thinks people are wasting time in meetings: he is a man who likes to get things done quickly and efficiently.

Penelope Bates

Penelope is the president of Bluenet's UK division. She has worked for Bluenet for 14 years and has wide experience in marketing. She is a clever negotiator who always seems to get exactly what she wants without losing her charm.

Penelope is quite a private person and she likes to keep her personal and business lives separate.

Jet Patel

Jet is a marketing executive working in Bluenet's UK office. He is extremely talented and enthusiastic, and always comes up with great ideas for marketing campaigns. He works in an efficient and organised way and is particularly good at making persuasive presentations. He handles other people well his colleagues find him a pleasure to work with.

BlueNet
Global Corporation

Part 1 Meet new partners

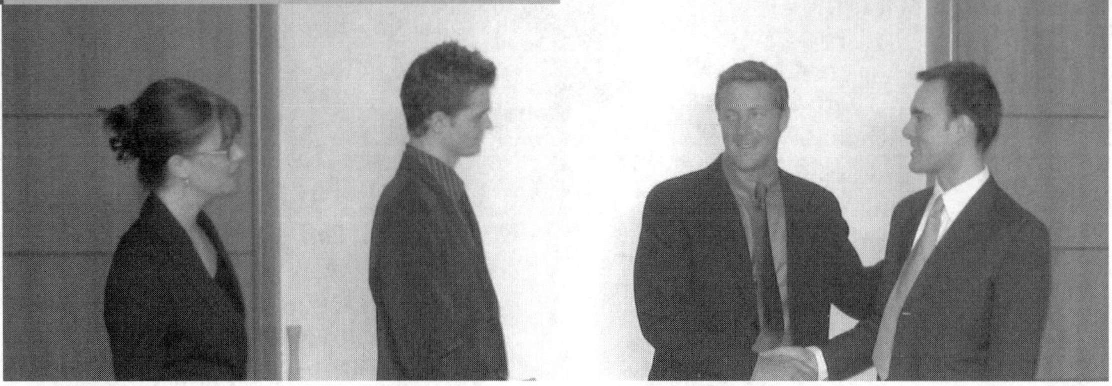

Preview **Introductions**

Unit 1 Match 1–4 with the appropriate response a–d.

1 I'm very pleased to meet you, Mr Fenno.
2 You must be Jet. It's good to meet you at last.
3 We're especially pleased that our CEO, Chuck, has flown in from New York just to be with us.
4 Perhaps you'd like to take us through today's agenda.

a Yes, I'm Jet, and it's a pleasure to meet you, too, Ms Morrison.
b Certainly, Chuck.
c 'Chuck', please. And I'm thrilled to meet you, Philippe.
d Thank you, Penelope. I'm really glad to be involved.

Video vocabulary

The words and expressions in the box are from the video. Use them to complete the gaps in the following sentences.

| crucial | corporate culture | agenda | internet penpals |
| global market | key | status symbol | star pupil |

1 The latest electronic gadget is a _____ for today's fashion-conscious teenagers.
2 Philippe and I became _____ about ten years ago and we used to send each other emails almost every day.
3 It isn't just important, it is absolutely _____ that we are ready in time for the international trade fair.
4 Low production costs are the _____ to achieving good profit margins.
5 So far, we have only sold our product locally, but we are hoping to access the _____ in the future.
6 Hilary worked very hard at college. In fact she was our _____ .
7 I left my job and went freelance to get away from office politics and the _____ . I'm much happier now.
8 There are several items on today's _____ , the most important of which is choosing a name for our new product.

First view

1 Before you watch Part one of the video, read these questions.

1. What are the names of the two companies in this video?
2. Which is the bigger company?
3. Who introduced Hilary to Jamie and Philippe?
4. Have Hilary and Jet met before?
5. Where does Chuck usually work?
6. What is the development code for Jamie and Philippe's invention and what is the name they eventually decide on?

Now watch Part one of the video and answer the questions above.

2 Watch Part one of the video again and choose the best endings for these sentences.

1. Jamie and Philippe met
   a through Jamie's father.
   b on the internet.
   c at the London Business School.

2. At the moment, Paridon sells
   a computers.
   b computer hardware.
   c computer software.

3. The FH1 is
   a relatively cheap to produce.
   b very expensive to produce.
   c likely to be expensively priced.

4. Paridon needs Bluenet's help
   a to finance the production costs of the FH1.
   b to provide technical support for developing the FH1.
   c to market the FH1 around the world.

5. The Paridon team are concerned
   a that Bluenet doesn't understand how the FH1 works.
   b that Bluenet will only target one market segment.
   c that Bluenet will not spend enough money on promotion.

6. Using a number in a product name is important because
   a it shows that you have more than one product to sell.
   b it gives an indication of the price of the product.
   c it suggests that this is the latest version of a product.

Part 1 ■ 5

Second view

➡ **Unit 1** **1** Watch section 0.01 to 5.44 again. Complete these diagrams of the Paridon and Bluenet company structures with the words in the box.

> Managing Director   Creative Director   Jamie Flacks
> CEO   President of UK division   Jet Patel

pariDON

Philippe Henetier   Hilary Morrison   1 _____
(2 _____)   (3 _____)   (Creative Director)
                                |
                          Support staff

Chuck Fenno   New York Headquarters
(1 _____)
                |
        Branches in 26 different contries
        |
   UK division

Penelope Bates
(2 _____)
        |
    3 _____
   (Marketing executive)
        |
    Marketing teams

➡ **Unit 2** **2** Watch section 5.45 to 9.33 again. Put a tick (✓) by the things that Jet has done to achieve his marketing objectives.

1 Make a workflow plan ☐
2 List the tasks that need to be done in each market segment ☐
3 Predict and try to prevent problems ☐
4 Allocate a certain amount of time to each task ☐
5 Break the tasks down into steps ☐
6 Draw up a detailed timetable ☐
7 Decide which tasks have the highest priority ☐

→ Unit 3

3 Watch section 9.34 to 12.56 again. Match these five pieces of advice about giving presentations to the excerpts from Jamie's presentation.

1 Make a structured plan for your presentation and tell the audience how your talk will be sequenced.
2 Make your objectives clear in the introduction.
3 Limit your presentation to three or four main points and signal these clearly.
4 Involve the audience by talking directly to them.
5 End with a strong summary.

a Firstly, let's think about who we're selling to ... Secondly, the name must tell the buyer what the product is ... OK, let's move on to our suggestions.

b First, I'd like to run through what we think are the most important ingredients that need to be in a name. So I'll start by listing the key points.

c So that's our top choice and that's our proposal. The iPlay 6010. It has everything. It has the key word in it, it really flows as spoken language, it tells you exactly what the product will do for you, and the '6010' makes it, straight away, two generations newer than the Nitron 40 – right from the start – and it's a great name for advertising slogans.

d ... just think what you can do with it ...

e We all know that we've got a great product. The question is, what are we going to call it?

After viewing

Unit 1

1 Complete these sentences with either the present simple or the present continuous form of the verb in brackets.

1 The Chicago trade fair (take place) _____ every year.
2 Jamie and Philippe (share) _____ an office at the moment because Philippe's office is being repainted.
3 Hilary makes sure that she, Jamie and Philippe (meet) _____ at least twice a week.
4 Jamie (prepare) _____ a PowerPoint presentation for tomorrow's meeting.
5 Jamie, Philippe and Hilary (meet) _____ the Bluenet team on Friday.
6 Jamie and Philippe (think) _____ that the FH1 will be the best product of its type on the market.
7 Hilary (feel) _____ sure that with Bluenet's help they can sell into the global market.
8 Chuck Fenno (stay) _____ in London for about three weeks.
9 Penelope (leave) _____ work early on Wednesday because she is going on a business trip to Brussels the next morning.
10 Hilary always (check) _____ that Jamie and Philippe are free before she arranges a meeting.

Unit 2

2 Complete this text with either a definite or indefinite article or with no article at all (Ø).

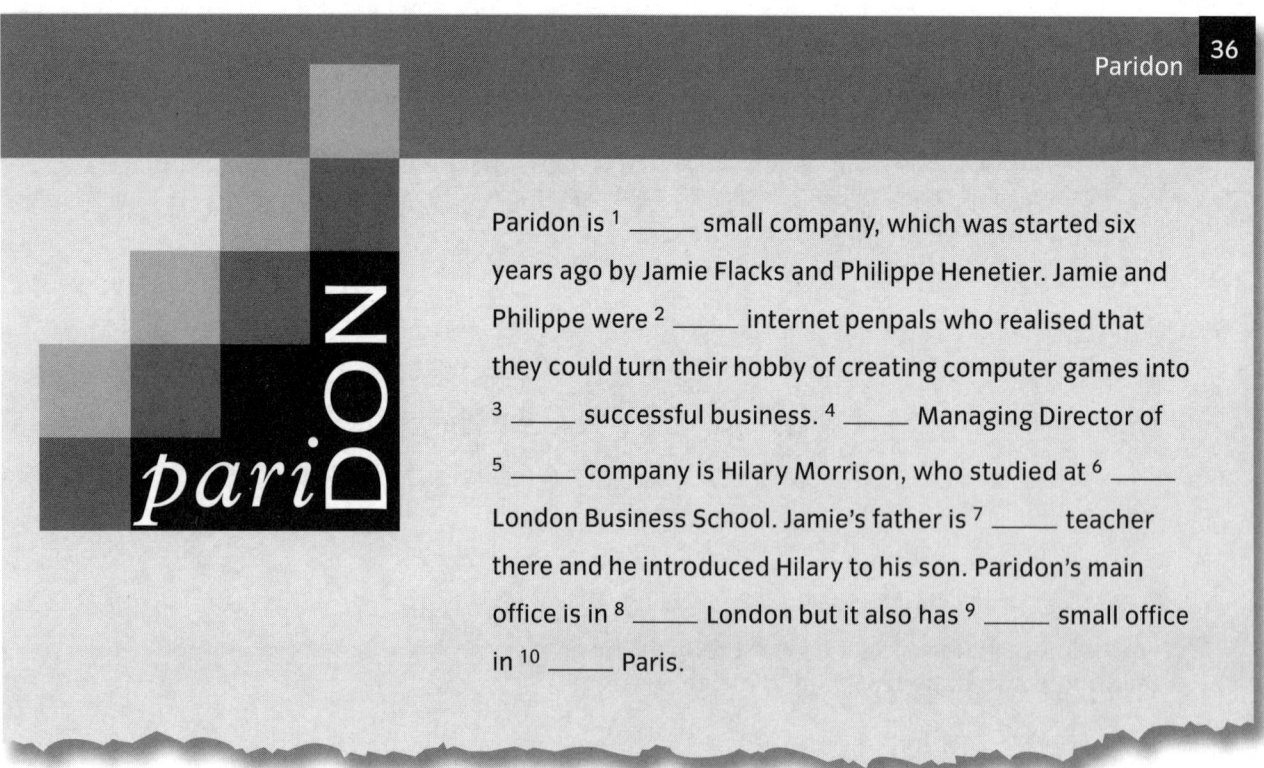

Paridon is ¹ _____ small company, which was started six years ago by Jamie Flacks and Philippe Henetier. Jamie and Philippe were ² _____ internet penpals who realised that they could turn their hobby of creating computer games into ³ _____ successful business. ⁴ _____ Managing Director of ⁵ _____ company is Hilary Morrison, who studied at ⁶ _____ London Business School. Jamie's father is ⁷ _____ teacher there and he introduced Hilary to his son. Paridon's main office is in ⁸ _____ London but it also has ⁹ _____ small office in ¹⁰ _____ Paris.

Unit 3

3 Complete the conversation with an appropriate future form, *going to*, *will* or the present continuous.

Hilary	Jamie, Philippe, don't forget that we (¹ meet) _____ the Bluenet team in the morning.
Jamie	Oh, right. The meeting's at their office, isn't it? (² go) _____ by train?
Hilary	Well, we (³ see) _____ them at 9:30 and the trains (⁴ be) _____ quite bad at that time in the morning. I think we'd better take a taxi.
Philippe	Right. I (⁵ call) _____ the taxi company and order a car for 8:30.
Hilary	Thanks, Philippe. Jamie, how (⁶ present) _____ the shortlist of names for the FH1?
Jamie	Well, I've prepared a PowerPoint presentation. First I (⁷ run) _____ through the most important ingredients for a name, and talk about who we (⁸ sell) _____ to. Then I (⁹ tell) _____ them the names we've thought of and explain why we think the iPlay 6010 is the best.
Hilary	That sounds good.
Jamie	I'm quite nervous about making this presentation.
Hilary	Oh, don't worry. I'm sure you (¹⁰ be) _____ fine!

Part 2 Deal with problems

> **Preview** **Giving presentations**
>
> **Unit 6** Look at this advice for people who are nervous about giving presentations. Tick (✓) the things that you think are good advice. Can you think of any other ways to stop being nervous when you have to speak in public?

1 Practise making your presentation to friends or colleagues first. ☐
2 Write down what you want to say beforehand and read it out. ☐
3 Don't look at your audience at all. Pretend that you are in an empty room or concentrate on the screen if you are using visuals. ☐
4 Make brief notes on your main points on small pieces of card. Refer to these when necessary. ☐
5 Before you begin, relax, take a deep breath and focus on the audience. Don't forget to smile. ☐

Video vocabulary

Look at the words and expressions in bold from the video. Match them with their definitions.

1 It's really important that we **evaluate** the meetings we've had with Bluenet.
2 He was very clear about the impact on the **bottom line**.
3 This is **make or break** now.
4 We must always keep our **objectives** in mind.
5 Innovation means **risk**.
6 We've **encountered** a problem with the Silicote-F.
7 We'll have to enforce the **default clauses** in our contract with you.
8 Each company has included the results of a **customer satisfaction survey** in its annual report.
9 It looks as if Crowne is the one to turn to when **the schedule is tight**.

Definitions

a sections of an agreement that say what compensation must be paid if one or other side fails to keep the agreement
b aims
c there isn't much time
d profit
e met, experienced
f assess
g questionnaire used to determine how satisfied customers are
h danger, possibility that something undesirable may happen
i the point at which we succeed or fail

First view

1 Watch Part two of the video and mark these statements *true* or *false*.

1 Jet has ignored Jamie's worry that there won't be enough people in the marketing teams. ☐
2 Increasing the number of people in the marketing teams will reduce Paridon's profits. ☐
3 Jamie and Philippe have agreed to hire some more support staff. ☐
4 Graham Kimberley can't deliver the screens for another three weeks. ☐
5 Hytek will have to pay some kind of compensation for failing to meet the agreed delivery date. ☐
6 One of Hytek's competitors has quoted a lower price for production of the screens. ☐
7 The bank has refused to lend Paridon any more money. ☐
8 It is Crowne's delivery record that convinces Jamie that they are the company to choose. ☐

2 Complete this summary.

Hilary, Jamie and Philippe get together to _evaluate_ the meetings they have had with Bluenet, to look at what they have ¹ _____ so far and to decide how they want to ² _____ .

Philippe believes that production of the iPlay is coming along ³ _____ until he gets a message from Graham Kimberley. When he phones him back, Graham tells him that they have encountered a ⁴ _____ with the Silicote-F and that the screens will not be ready in time in the ⁵ _____ that Paridon wants. Paridon will have to find another company to produce the screens.

Hilary persuades the bank to lend Paridon some extra money. However, Jamie has to convince them that it won't go wrong again. He ⁶ _____ his presentation in front of his colleagues.

Jamie has looked at two companies that could supply the screens. He believes they won't now be as ⁷ _____ as when they first quoted for the work. Both companies have similar market share and both have included the results of a customer ⁸ _____ survey in their annual reports. He recommends Crowne because they have a good record for ⁹ _____ and don't seem to be worried by ¹⁰ _____ schedules or changes of schedule.

Second view

Unit 4

1 Match the phrases Hilary uses a–g with the functions 1–7.

1 Signal the start of a meeting
2 Greet and welcome the participants
3 Explain the background to the meeting
4 Ask for a first contribution
5 Respond to a contribution and elicit one from someone else
6 Summarise and outline future action
7 Signal a return to social conversation

a Philippe, what do you think the key outcomes of the first meeting were?
b Let's sum up and move forward. What we need to do is focus on building a really good relationship with Bluenet.
c Talking of progress, how is the iPlay coming along?
d Excellent. And Jamie, how was the meeting with Jet?
e OK, let's start.
f Thanks for coming along, guys; I know how busy you are.
g I think it's really important that we evaluate the meetings we've had with Bluenet ... look at what we've achieved so far ... and decide how we want to progress.

Unit 5

2 Watch Section 16.17 to 18.07 again. Complete this phone message, which Philippe found on his desk when he returned from his meeting with Hilary and Jamie.

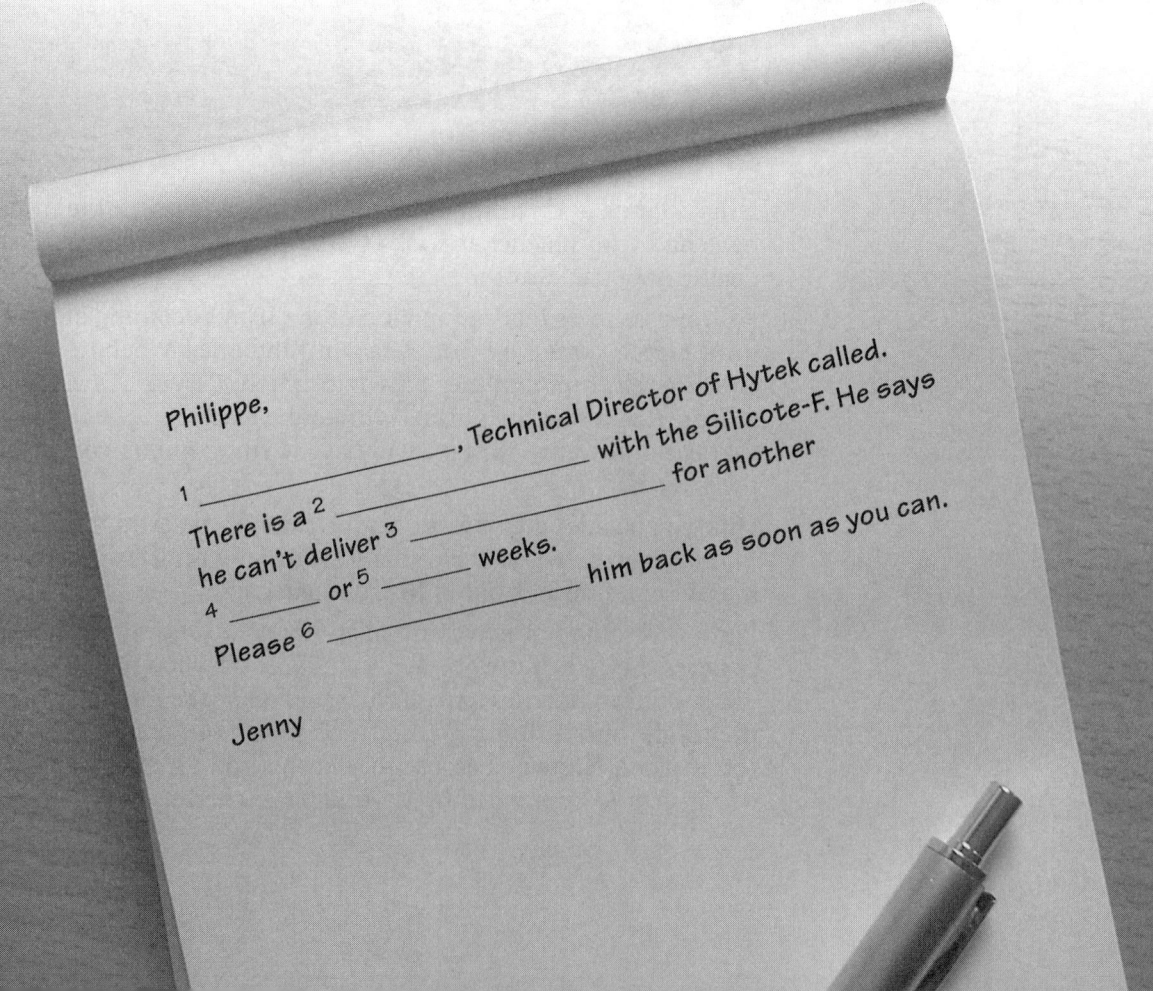

Philippe,

1 _____, Technical Director of Hytek called. There is a 2 _____ with the Silicote-F. He says he can't deliver 3 _____ for another 4 _____ or 5 _____ weeks. Please 6 _____ him back as soon as you can.

Jenny

→ Unit 5 **3** Watch Section 16.17 to 18.07 again. Match expressions 1–5 with the functions a–e.

1 Can I just clarify what you said?
2 We'll have to use another supplier for the screens now, and we'll have to enforce the default clauses in our contract with you.
3 This means that Teegan will beat us.
4 We've encountered a problem with the Silicote-F.
5 I'm sorry.

a Explain a problem
b Ask for confirmation of a message
c Outline action that will be taken
d Apologise
e Predict the consequences of a problem

→ Unit 6 **4** Watch Section 18.08 to 21.58 again. Number these visuals in the order that Jamie talks about them.

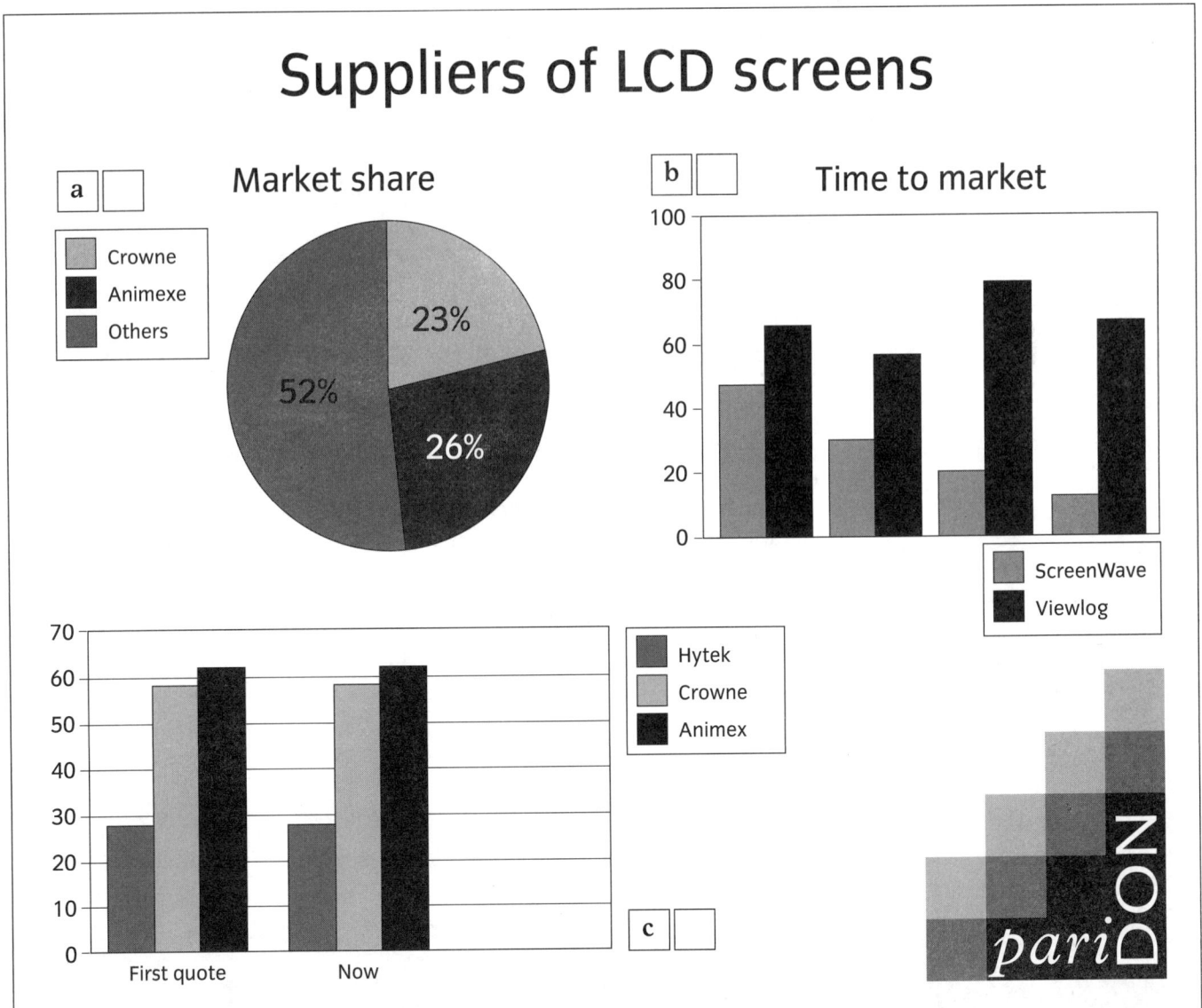

5 Match these expressions with the things Jamie is describing. Choose from a–f; you will need to use one twice.

1 no significant difference there
2 again, they were very similar
3 slightly in front
4 came onto the market in the shortest time
5 the one to turn to when the schedule is tight

a Crowne and Animex's prices
b Crowne and Animex's customer satisfaction surveys
c Crowne's customer satisfaction survey
d Four products produced with parts or processes supplied by Crowne
e Crowne
f Animex
g Crowne and Animex's market share

After viewing

→ Unit 4

1 Complete these sentences with either the past simple or the present perfect form of the verbs in brackets.

1 I think it's really important to look at what we (achieve) _____ so far.
2 Graham Kimberley (encounter) _____ a problem with the Silicote-F and he doesn't think he can deliver the screens on time.
3 Jamie says that Bluenet certainly (put) _____ good people on the project and he believes that now they (got) _____ every market segment covered.
4 Jet (pass) _____ on the message that Jamie and Philippe were worried about the number of people allocated to each team and as a result, Chuck (phone) _____ Hilary yesterday to talk about this.
5 During their conversation Chuck (agree) _____ to increase the numbers of people in the teams for each market segment.
6 Jamie believes that he and Philippe (improve) _____ the understanding that Bluenet has of the iPlay and that they will now be able to market it more efficiently.

Unit 5

2 Choose the appropriate modal forms to complete these sentences.

1. If Paridon don't solve the problem with the screens, they _____ be able to launch at Chicago.
   a might	b won't	c shouldn't

2. Finding another screen supplier _____ add to Paridon's costs.
   a can	b shouldn't	c will

3. The customer satisfaction surveys provided by Crowne and Animex _____ be reliable because they were verified by the auditors.
   a can	b might	c should

4. Crowne _____ be a good company because its customers are very happy with the service it provides.
   a would	b must	c could

5. The difference in price between Hytek and Crowne _____ be as great as when they first quoted for the job.
   a mightn't	b needn't	c mustn't

6. Buying screens from Crowne _____ cost them more, but Jamie is confident that Crowne will deliver on time.
   a won't	b may	c can

Unit 6

3 Complete these sentences with the correct comparative or superlative forms of the adjectives in brackets.

1. When Hytek, Crowne and Animex were first asked for quotes for the screens, Hytek's quote was (low) _____ .

2. Jamie looked at the four products that came onto the market in (short) _____ time.

3. Crowne seem to be slightly (good) _____ than Animex when it comes to customer satisfaction.

4. Of the three companies he looked at, Crowne seem to have (good) _____ record for delivery.

5. Crowne are (expensive) _____ than the other companies but they can be relied on to deliver on time.

6. Crowne would certainly be (reliable) _____ than Hytek.

Part 3 Find solutions

Preview **Brainstorming**
Unit 7

In the next part of the video you will see the teams taking part in a brainstorming session. They are trying to find the solution to another problem that has arisen over the launch of the iPlay 6010.

Brainstorming can be a very productive method for getting ideas – often ideas that would not have developed in other circumstances. Apart from sometimes producing really valuable ideas, brainstorming can be great fun and a terrific team-building activity. However, it's important to run brainstorming sessions properly.

Complete these suggestions for running brainstorming sessions with the words in the box.

| problem | growing | shortlist | atmosphere | sensitive |
| criticise | participants | forgotten | participation | valued |

1 Keep the number of _____ down to six or so.
2 Start by defining the _____ very clearly.
3 Never _____ suggestions.
4 Keep the _____ relaxed at all times and encourage excitement and _____ .
5 Make *every* contributor feel _____ .
6 In multicultural meetings, always be _____ to different attitudes towards what is the proper degree of participation.
7 Write everything down so that people can actually *see* that the number of ideas is _____ , and what they are, and so that nothing gets _____ .
8 _____ the ideas and evaluate them *after* the session, rather than interrupting the flow.

Video vocabulary

1 Match the words in column A with those in column B to make common collocations.

A		B	
1	aesthetic	a	adopters
2	early	b	campaign
3	marketing	c	pricing
4	retail	d	appeal
5	global	e	strategy
6	advertising	f	buyers
7	spoiler	g	scale
8	competitive	h	outlets
9	potential	i	disaster

2 Now use the collocations to complete these sentences.

1 Teegan has its own _____ in which it can sell its products.
2 _____ are the people who are the first to show an interest in and buy a new product.
3 Paridon and Bluenet need to achieve maximum impact on a _____ with minimal cost to achieve their aims.
4 Teegan are planning a massive _____ with TV commercials and newspaper articles.
5 Our competitors launched a month earlier than us and their product was better and cheaper than ours. For us this was a _____ .
6 People used to value practicality and durability above design, but now _____ may be the most important factor in the marketing of electronic goods.
7 A _____ such as launching earlier than your competitors is designed to damage their sales potential.
8 Our research shows that there are _____ in all market segments.
9 _____ is an attempt to gain market share by making sure a product costs less than rival products.

First view

1 Before you watch Part three of the video, read these questions.

1 What is the problem that the Paridon and Bluenet teams are meeting to discuss?
2 When will people be able to buy the Nitron 40?
3 Why don't they want to launch the iPlay 6010 before Chicago?
4 Who will be invited to the special preview events?
5 Where will the preview events be held?
6 Why does Jet think it is important that the guests at these events are not all male?
7 What is the most important consideration nowadays when people buy domestic technological devices?
8 What do Jamie, Philippe and Chuck talk about over dinner?

Now watch Part three of the video and answer the questions above.

2 Mark these statements *true* or *false*.

1. Paridon and Bluenet are worried that early adopters will buy the Nitron 40 even if they know that a better product is coming onto the market soon. ☐
2. The Nitron 40 will be available in Teegan's retail outlets as soon as their commercials start to air. ☐
3. Bluenet are planning an advertising campaign to take place before the Chicago trade fair. ☐
4. Paridon and Bluenet want to make their preview events exclusive. ☐
5. The Bluenet representatives knew in advance why they had been summoned to London. ☐
6. Jet presents information on changing trends in the purchase of domestic technological devices. ☐
7. There have been few changes in the last 25 years in the type of people who buy domestic technological devices and the reasons why they do so. ☐
8. Jet has never visited Paris. ☐

Second view

→ Unit 7

1 Watch section 22.09 to 27.05 again. Tick (✓) the ideas which are mentioned in the brainstorming session.

1. Setting up a trade fair of their own. ☐
2. Having an advertising campaign that compares the iPlay 6010 with the Nitron 40. ☐
3. Selling the iPlay 6010 at special preview events. ☐
4. Allowing certain people to see and try the iPlay before the Chicago trade fair. ☐
5. Setting up a network of retail outlets. ☐

Which of these ideas is eventually accepted?

2 Watch section 23.20 to 27.05 again. Tick (✓) the ways of making suggestions that you hear.

1. Why don't we ...? ☐
2. How about ...? ☐
3. If I were you ... ☐
4. I suggest ... ☐
5. What if we ...? ☐
6. Couldn't we ...? ☐
7. What would happen if we ...? ☐
8. I think we should ... ☐
9. I suggest we should ... ☐
10. We might be better off ... ☐

Unit 8 **3** Watch section 27.08 to 31.52 again. Which of these things does Jet do to get the attention of his audience?

1 He starts his presentation with a dramatic statement. ☐
2 He often uses the word *you* to make what he says relevant to them. ☐
3 He raises his voice so that they can't avoid hearing him. ☐
4 He maintains eye contact with them as he speaks. ☐
5 He waves his arms around a lot to attract their attention. ☐
6 He refers to particular members of the audience by name. ☐
7 He uses visuals to illustrate the more technical parts of his presentation. ☐
8 He points at particular members of the audience so they know that he is watching them. ☐

Unit 8 **4** Watch section 30.40 to 31.52 again. Tick (✓) the expressions that Jet uses to talk about the visuals in his presentation.

1 Take a look at this graph ... ☐
2 As you can see ... ☐
3 You can see that ... ☐
4 This graph shows ... ☐
5 You'll notice that ... ☐

5 Which of these topics do you think would be suitable for conversation at a dinner with business colleagues? Tick (✓) those that you think would always be suitable, put a question mark (?) next to those that might be suitable in certain conditions and put a cross (✗) next to those that you think would never be suitable.

1 The weather ☐
2 Politics ☐
3 Personal relationships ☐
4 Sport ☐
5 Religion ☐
6 Food ☐
7 Your surroundings ☐
8 Your childhood/teenage years ☐

6 Watch section 31.53 to 33.18 again. Which of the topics in Exercise 5 are asked about or talked about? Does anyone show signs of being uncomfortable about anything they are asked?

Part 3 ■ 19

After viewing

Unit 7

1 Match the two halves of the sentences.

1 If Paridon had their own retail outlets,
2 If they make their preview events really exclusive,
3 If early adopters see that a product from a respected brand is on the market,
4 If they invited everybody to the preview events,
5 If they decided to have an advertising campaign,
6 If people can actually see and try an iPlay 6010,

a they would have to spend a lot of money to make it as good as Teegan's.
b they may be prepared to wait until it comes onto the market.
c they will create a cult straightaway.
d they will buy it even if they know that a better one is on the way.
e they would be able to sell more units of the iPlay 6010.
f they would appear less exclusive.

Unit 8

2 Rewrite each sentence as in the example, keeping the meaning the same.

There was a dramatic fall in profits last year.
Profits fell dramatically last year.

1 There was a sharp increase in the number of women buying technological devices.

2 There was a slight fluctuation in the sales of the iPlay 6010 over a three-month period.

3 There was a slight drop in sales in July.

4 There was a steady rise in prices in 2005.

5 There was a gradual decline in interest in the Nitron 40 after the initial excitement.

6 There was a striking similarity between sales of the Nitron 40 and the iPlay 6010 in the first two months.

→ Unit 9 **3** Complete these sentences with appropriate relative pronouns. Put a tick (✓) if the clauses are defining.

1 Using a new supplier for the screens will cost an extra seven per cent, _____ is a lot less than they had feared. ☐

2 Teegan is launching the Nitron 40, _____ is very similar to the iPlay 6010, two months earlier than they had planned. ☐

3 Paridon want to launch at Chicago because it's the place _____ they would get most impact. ☐

4 Bluenet will hold the special preview events in every city _____ they have offices. ☐

5 The bosses of all the different Bluenet offices are the people _____ will have to organise the previews. ☐

6 The guests _____ are invited to the previews will be key industry correspondents and consumers. ☐

7 They can try an iPlay at the previews, but if they want to buy one, they will have to wait till after Chicago, _____ they will be on sale in the shops. ☐

8 Twenty-five years ago, the people _____ bought new domestic technological devised were 74 per cent male. ☐

Part 4 Reach agreement

Preview **Presenting a structured argument**

→ Unit 10 Match statements 1–6 with their more polite equivalents a–f.

1 Your plans are no good.
2 Our ideas are better.
3 We're going to tell you why.
4 There's nothing wrong with our plans; what do you mean?
5 Your idea is very strange.
6 Your strategy is much too dangerous.

a We appreciate that it's an interesting idea to …
b We think we've got the correct formula, but let's hear what you have to say.
c We're worried about your plans.
d We'd like to go through the reasons with you.
e We really do think that it's too risky an approach.
f We think that another approach would be more successful.

Video vocabulary

1 For each verb in column A, circle the noun in column B that doesn't collocate with it.

	A	B			
1	resolve	a disagreement	an argument	a problem	an approach
2	change	a plan	an opinion	a problem	a decision
3	propose	a problem	a strategy	an approach	a solution
4	support	a decision	a disagreement	a team	an argument
5	raise	an issue	a question	a point	a strategy
6	arrange	a meeting	a negotiation	a private viewing	a conference

2 Put the words in the box in the appropriate sections of the word map. One word can be used twice because it has two meanings.

argument	strategy	decision	proposal	issue

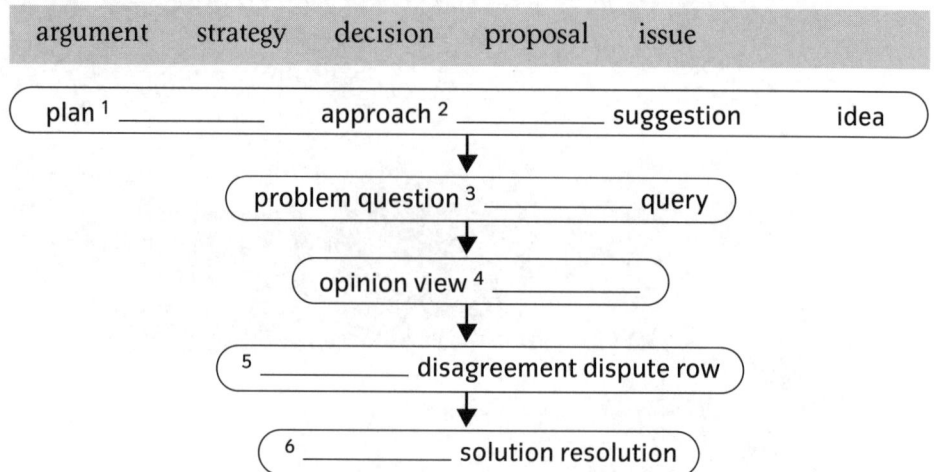

First view

1 In Part four of the video, you will see a meeting between the Paridon and Bluenet teams. Before you watch the video, read these extracts from their discussion. Mark them *P* or *B* according to which team you think says them.

1 We're worried about your plans for previewing the iPlay package. ☐
2 We think that another approach will be more successful. ☐
3 We think we've got the correct formula. ☐
4 Well, we don't think it's such a good idea. ☐
5 We understand that there are very good reasons for your strategy, but we really do think that it's too risky an approach. ☐
6 We'd like to go through the reasons with you. ☐
7 But what you're suggesting is going to add a great deal to our costs. ☐
8 But we still need to make sure that the guests at the previews feel it's exclusive. ☐

Now watch Part four of the video and see if you were right.

2 Watch Part four of the video again and choose the best endings for these sentences.

1 The Bluenet team's original plan for the marketing materials is
   a to have them look slightly unfinished.
   b to make them the best possible quality.
   c not to have any at all.

2 The Bluenet team agrees to produce better quality materials but they are still concerned
   a that they will cost too much.
   b that they really ought to be producing commercials.
   c that the guests need to feel that the preview events are exclusive.

3 Bluenet may have to pay more than usual for the printing of the materials because
   a Penelope isn't very good at negotiating with printing companies.
   b the printers could take advantage of the fact that they have little time to produce them.
   c they don't usually produce high-quality materials for small clients.

Part 4 ■ 23

4 Geoff Byrne will find it difficult to reduce the printing costs because
  a he will need to hire more staff to get the job done on time.
  b cheaper materials are not available at the moment.
  c he is busy with two other jobs for Bluenet.

5 Hilary's suggestion is to
  a guarantee to give two more jobs to Geoff's company so that the cost of the materials and the extra staff can be spread.
  b lend Geoff some of the Bluenet staff to help get the work done on time.
  c cancel the order for the holiday company so that more of Geoff's staff are free to work on the Paridon materials.

6 The outcome of Hilary's meeting with Geoff Byrne is
  a satisfactory to both parties.
  b satisfactory to neither party.
  c satisfactory to Hilary but not to Geoff.

7 Hilary asks for the pre-launch meeting to be held at the Paridon offices because
  a the Bluenet office is being prepared for the presentation the next day.
  b the Paridon team need to be there to deal with any problems that arise with their staff.
  c she needs to receive phone calls from people who are upset at not being invited.

8 The teams decide to avoid upsetting valuable customers by
  a inviting everyone who contacts them to come to one of the preview events.
  b getting a larger venue in one or two cities so they can invite more people.
  c sending literature to people who haven't been invited and inviting some of them to private viewings.

Second view

> Unit 10

1 Watch section 33.29 to 37.43 again. Tick (✓) the reasons Jamie and Philippe give for producing better quality marketing materials.

1 If the marketing materials aren't good quality, the guests won't feel it is a privilege to be there. ☐
2 The materials will be shown to people who aren't at the previews, so they need to be good. ☐
3 Well-designed materials will make the whole concept of the iPlay more attractive. ☐
4 The materials will be competing with the commercials for the Nitron 40. ☐
5 If the materials show the iPlay in use, it will make people think how they would use it themselves. ☐
6 The screens are now cheaper than they expected so they have money to spare. ☐

→ Unit 11 **2** Watch section 37.44 to 41.10 again. Tick (✓) the strategies that Penelope uses to get what she wants in her negotiation with Geoff Byrne.

1 She makes a practical suggestion for a solution which can benefit both parties. ☐
2 She tells Geoff exactly how much Bluenet is prepared to pay for the printing. ☐
3 She makes a subtle threat that Byrne Graphics may lose Bluenet's custom for future projects. ☐
4 She makes Geoff feel that she personally is on his side. ☐
5 She presents using another company in future as a problem for Bluenet when in fact it is a problem for Byrne Graphics. ☐
6 She names the other companies who have quoted lower than Byrne Graphics. ☐
7 She asks for a quick decision and confirmation in writing. ☐

Now match the strategies you have ticked with these excerpts from the video.

a I keep saying that the quality you offer is worth paying for ...
b We really need to have a decision soon. When can you get back to me? ... Can you get it to me in writing? Fax or e-mail?
c What if we confirm the other two jobs that you've quoted for recently, the car launch and the holiday company, would that help you? You could get staff in for three jobs instead of one, and get a better price on the stock for a much larger order?
d And that might cause us problems in the future, if we want to come back to you when we've started building a relationship with another supplier.
e I'm coming under pressure to use someone less expensive.

➲ Unit 12 **3** Watch section 41.14 to 44.09 again.

Match each excerpt from the video 1–10 with its function a–j.

1 Penelope, is there anything you would like to raise?
2 Can I say something?
3 But don't you think that ...
4 Can I just finish?
5 Jet, you wanted to say something?
6 Go on, Jamie.
7 That's the reaction I've had too.
8 Erm ... I suppose so.
9 What do you think, Hilary?
10 Is everyone happy with that?

a Acknowledge that someone has a point
b Acknowledge that someone else has indicated that they wish to speak
c Invite someone else to give their opinion
d Interrupt
e Invite someone to start the discussion
f Get your turn to speak
g Encourage someone to expand on what they are saying
h Resist an interruption by someone else
i Check that all participants are in agreement
j Support someone else's point

After viewing

➲ Unit 10 **1** Match the two halves of the sentences.

1 If people see beautifully produced marketing materials as well as a stunning unit,
2 Even if Paridon and Bluenet doubled their marketing budget,
3 That might cause us problems in the future,
4 If we aren't very careful,
5 If we confirm the other two jobs that you've quoted for recently,
6 If Geoff Byrne hadn't agreed to reduce the cost of the printing,

a if we want to come back to you when we've started building a relationship with another supplier.
b Penelope would have used a different company to produce the marketing materials.
c you could hire extra staff for three jobs instead of one, and get a better price on the stock.
d they would still spend less than Teegan is spending on their advertising campaign.
e we could upset a lot of valuable customers and our whole plan will backfire.
f they'll be even more impressed by the iPlay.

Unit 11

2 Complete the sentences with the words in the box.

| to get | printing | to do | building | to resolve | having | to think |
| paying | to start | previewing | to use | saying | to make | |

1 There's a disagreement that the teams need _____ immediately.
2 We're worried about your plans for _____ the iPlay package
3 We can understand why you want _____ it the way you've proposed.
4 This is meant _____ them feel that they are being let into a secret.
5 They will start _____ of themselves as owners.
6 We're trying _____ a new brand.
7 It may be difficult _____ everything done in time.
8 We're all a little surprised at the quote you've given us for _____ the re-worked visuals.
9 I keep _____ that the quality you offer is worth _____ for.
10 I'm coming under pressure _____ someone less expensive.
11 That might cause us problems in the future, if we want to come back to you when we've started _____ a relationship with another supplier.
12 I hope you don't mind _____ this meeting here.

Unit 12

3 Complete the fax to Penelope with the words in the box.

| need | have to | need to | has to | ought | must | mustn't |
| should (x2) | doesn't have to | | | | | |

To: Penelope Bates, Bluenet   From: Geoff Byrne
Re: Revised quote for printing work Date: February 21st

Dear Penelope
Thank you very much for seeing me this morning. I realise that you ¹ _____ a quick decision on the revised quote for printing the Paridon materials, so I have done some calculations and hope that you will be satisfied with the result.

In our meeting, we agreed the following:
- We ² _____ use the best quality paper for these materials.
- Byrne Graphics will ³ _____ hire extra staff to complete the printing in time.
- The cost of this ⁴ _____ be spread over three projects, Paridon, Delux Cars and Getaway, all of which you guarantee to place with Byrne Graphics.

On the basis of this, I now think that we ⁵ _____ to be able to reduce our quote by seven and a half per cent This ⁶ _____ give you the materials you want at a price that is acceptable to you.
Please note that in order to get the printing on time:
- You ⁷ _____ supply the full text and all the photos by next Monday at the latest.
- You ⁸ _____ ask for late changes to the text.

Also please note that someone from your company ⁹ _____ be available to check the first proof of the materials on Thursday morning. This person ¹⁰ _____ come to Byrne Graphics to do this, we can courier a copy of the proofs to your office on Wednesday evening.

I hope that you will find this revised quote acceptable and that we can continue to do business together in the future.

With all best wishes

Geoff

Geoff Byrne

Part 5 Celebrate success

Preview **Summarising**

Unit 13 Complete the gaps in this summary of the story so far.

The teams from Paridon and Bluenet are working together to ¹ _____ the iPlay 6010, a new combined music player, games machine and internet phone ² _____ by Jamie Flacks and Philippe Henetier. So far, they have encountered two major ³ _____ . The first was that the company making their screens let them down and they had to find another ⁴ _____ . They have now done this with only a slight increase in costs. The second was that a rival company, Teegan, decided to launch its own product, similar to the iPlay, ⁵ _____ months earlier than planned. The teams have countered this problem by holding special ⁶ _____ events in Bluenet offices around the world at which key industry correspondents and ⁷ _____ can see and try an iPlay before it is launched at the Chicago trade fair. This will mean that excitement about the iPlay will be created before it is actually launched and key people will be able to handle and use the iPlay while they can only see Teegan's product, the Nitron 40, in ⁸ _____ .

Video vocabulary

1 Match the words in column A with their equivalents in column B.

A	B
1 confidential	a end
2 modify	b practice
3 rehearsal	c position
4 conclusion	d best
5 location	e secret
6 premier	f alter

2 Choose the best explanation for each of these sentences.

1 'We gave it our best shot.'
   a We did the best that we could.
   b We only just missed the target.
   c We supplied a superior brand of whisky.

2 'I'd like to clear up any loose ends.'
   a The floor needs to be swept clean.
   b There is some unfinished business.
   c The standard of manufacture is not high enough.

3 'We need to be finished by eight sharp.'
   a We want to be out of here by eight o'clock exactly.
   b The knives need to be sharpened by eight o'clock.
   c The finish on our knives is not sufficiently sharp.

4 'I'm speaking off the record.'
   a   The cassette recorder has not been turned on.
   b   I'm just quoting what someone else has said.
   c   I don't want you to broadcast what I am saying.

5 'We didn't quite catch that.'
   a   We didn't hear what you said.
   b   We just missed that train.
   c   That moved too fast for us.

6 'The industry wires were buzzing around the world.'
   a   There was a malfunction in the international telephone system.
   b   People were very excited about what they had heard.
   c   Lots of people around the world were making long telephone calls.

7 'This was the icing on the cake.'
   a   This was the thing that spoiled the celebration.
   b   This was something good which made the previous bad news more acceptable.
   c   This was a bonus on top of what had already been achieved.

First view

1 Watch Part five of the video and mark these sentences *true* or *false*.

1 Penelope wants the hotel staff to check the invitation passes. ☐
2 Philippe thinks he and Jamie ought to do the meeting and greeting at the door. ☐
3 Jet devoted part of his presentation to explaining why the iPlay was better than the Nitron. ☐
4 Jamie and Philippe are reluctant to reveal their future plans. ☐
5 There won't be an advertising campaign for the iPlay 6010. ☐
6 Feedback from the other Bluenet branches was very good. ☐
7 Penelope persuaded the organisers of the Chicago trade fair to improve the position of the Paridon stand. ☐
8 The next model of the iPlay will be the iPlay 6020. ☐

2 Watch Part five of the video again and number these events in the order in which they happened.

a   The teams celebrated the success of the launch. ☐
b   The teams met to discuss their strategy for the day. ☐
c   Penelope telephoned Chicago to negotiate a better stand at the trade fair. ☐
d   Jet answered questions from the audience. ☐
e   The launches took place in 80 per cent of Bluenet's offices around the world. ☐
f   Jet gave his presentation at the launch in London. ☐
g   Penelope raised the problem of security arrangements. ☐

Second view

Unit 13

1 Watch Section 44.20 to 47.33 again. Do you agree with this assessment of Chuck?

Chuck is very experienced at running meetings. Some people work more flexibly, but Chuck likes to keep firmly to an agenda. He always starts on time and he keeps to time limits – with the discussion sharply focused on the agenda items. However, he knows that it's really important to listen carefully because someone may bring up an important and relevant point that isn't on the agenda. Chuck doesn't really like his plan being altered, so anyone who wants to interrupt has to be quite forceful.

2 Choose the best word to complete each of these sentences.

1 The final team _____ is taking place, just to clear up any loose ends.
   a briefcase   b breathing   c briefing   d breathless

2 Our _____ this morning is to check that everything is organised and that everyone knows exactly what's going to happen
   a objection   b objective   c obligation   d objecting

3 You all have a copy of the _____ , I hope?
   a addenda   b agenda   c avenger   d gender

4 My plan is to _____ any necessary actions and get us away at eight.
   a assassinate   b assign   c assist   d assert

5 We need to get them out of their seats and _____ as early as possible.
   a rotating   b spinning   c spiralling   d circulating

6 This is a very commercially _____ event.
   a sensible   b sensual   c sensitive   d sentient

7 At least one of us should _____ the guests as they arrive.
   a greet   b grope   c grieve   d guess

8 Hilary must be in the reception room with time to talk to people _____ .
   a independently   b indescribably   c individually   d indecisively

30 ■ Part 5

Unit 14

3 Watch section 47.35 to 51.46 again. Choose the best answers to these questions. For some questions there may be more than one answer.

1. Why did Jet wait till the end of his presentation to tell the guests they would have a chance to try the iPlay?
   a. Because it would come as a surprise and generate excitement.
   b. Because he wasn't sure whether the units would arrive on time.
   c. Because he thought they would lose concentration if he told them at the beginning.

2. Why didn't Jet mention the Nitron 40 until the very end of his presentation?
   a. He wanted the focus of attention to be the iPlay 6010.
   b. Talking about the Nitron 40 in detail would give Teegan free publicity.
   c. He hoped that the guests would know nothing about the Nitron 40.

3. Why was Jenny Michaels so anxious to hear about Paridon's future plans?
   a. She thought the iPlay wasn't good enough to beat the Nitron 40 and hoped they had a better model coming up.
   b. She thought it would be easier to make customers wait for the iPlay 6010 to be launched if she could tell them about Paridon's future developments.
   c. It would be easier to sell two products rather than one and she hoped the iPlay 7000 would be ready at the same time as the 6010.

4. Why did Robin Davis want to know the details of the advertising campaign?
   a. He didn't believe that there really was going to be an advertising campaign.
   b. He wanted to be consulted on the style of the campaign.
   c. He didn't want to appear ignorant if customers asked him about it.

4 Answer these questions.

1. What were the two main questions that Jet was asked?
2. Which of the questions didn't he hear?
3. Which of the questions didn't he understand?
4. Which one of them did he answer?
5. Who answered the other question?

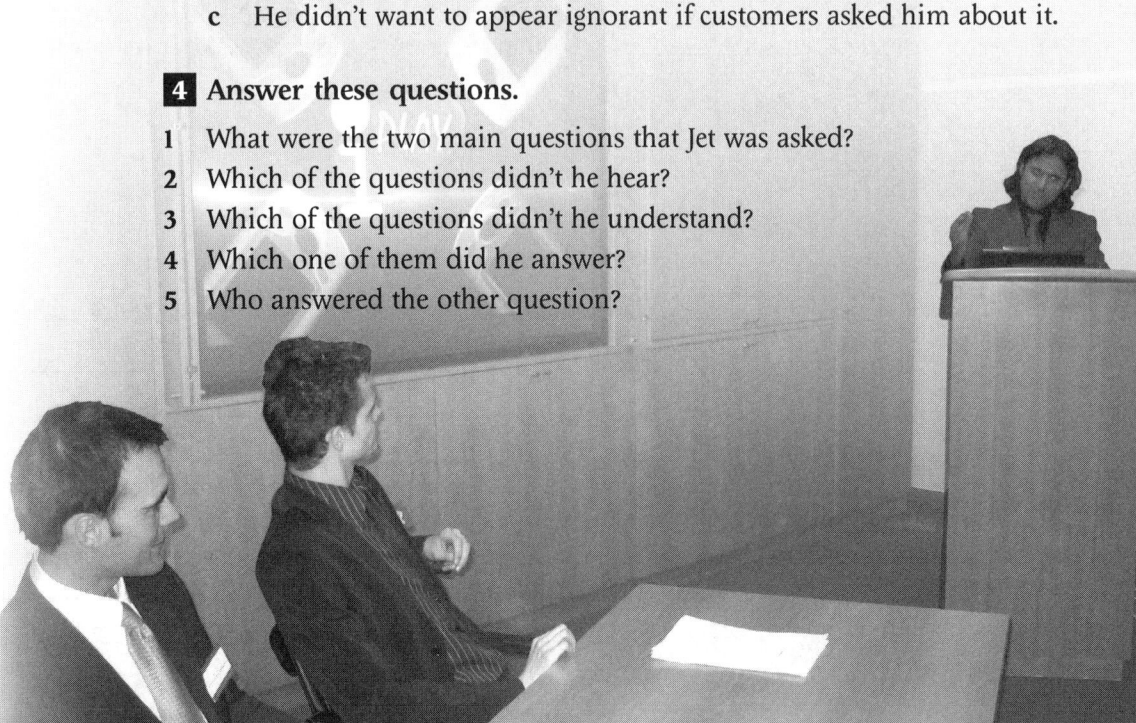

➡ Unit 15 **5** Watch section 51.47 to 54.16 again.

Match the excerpts from the video a–g with functions 1–4. There are two excerpts for some functions.

1 Assess the success of the day
2 Show appreciation
3 Praise individuals for their contribution
4 Respond to praise

a And can I just say, Chuck, how much Philippe and I appreciate the faith you've shown in us?
b I don't know what to say! I was just part of a team.
c I'd just like to say fantastic team, fantastic product!
d And a special thanks to you, Philippe.
e A brilliant job, everyone – especially you guys from Bluenet.
f It's been great working together.
g What a first class day.

After viewing

[reported speech, passives; past modals]

➡ Unit 13 **1** Imagine you were at the final meeting before the London launch. Use reported speech to tell a colleague what Chuck said.

> We are on schedule, but there's no time for any errors. The rehearsal of the presentation will take place ten minutes earlier than planned because the venue staff need time to vacuum the room *after* the rehearsal, not before, which we weren't expecting. That's at 9:45. After the rehearsal, we'll have a feedback session. That will give the event production team a chance to do a final check. And we'll need the Audio Visual supervisor in here for that meeting. Jet, please make sure he knows.

Begin like this:

Chuck said that we were on schedule, but there was no time for any errors.

He explained that ....

→ Unit 14 **2** Rewrite the following sentences, putting the verbs in italics in the passive. Do not include the agent if it is not important.

1 They *agreed* that Jet wouldn't even mention the Nitron 40 in the main part of his presentation.
   _____

2 The moment we *finalise* those plans, we'll be in touch with you all.
   _____

3 By mid-afternoon they had *held* the launches in 80 per cent of Bluenet's offices.
   _____

4 Penelope *made* the decision to call Chicago and get a better stand at the fair.
   _____

5 Now that we have *launched* the iPlay, let's go out to celebrate.
   _____

→ Unit 15 **3** Read the following article, which appeared in *Futuretech* magazine. Complete the statements below with the appropriate past modal forms of the verbs in the box.

| price | challenge | design | include | make | spend |

Teegan loses out to Paridon

Shares in technology firm Teegan fell sharply yesterday when it was revealed that sales of their Nitron 40 had been disappointing. The main factor in the failure of the Nitron has been the success of rival company Paridon's iPlay 6010.

Both products are a combination of music player, games machine and mobile phone, but the iPlay has many more product features than the Nitron: it is smarter looking, lighter, easier to use and cheaper.

Key industry correspondents and consumers were given a chance to try the iPlay 6010 before it was launched, while Teegan relied on their advertising campaign to generate interest. Despite this huge advertising campaign, the Nitron 40 has failed to grab the attention of the lead market segment, the 20-somethings, in the way that the iPlay 6010 has done. Industry analysts point out that the design of the Nitron 40 is just too old-fashioned and the unit too bulky and heavy to appeal to fashion-conscious young professionals.

In the meantime, the Paridon product is going from strength to strength and the market eagerly anticipates the launch of their new model, the iPlay 7000.

1 Teegan _____ their product with the 20-something market segment in mind.
2 Teegan _____ the Nitron 40 smaller, lighter and easier to use.
3 Teegan _____ more product features in the Nitron 40.
4 Teegan _____ less money on advertising and more on design.
5 Possibly if Teegan had given customers a chance to try the Nitron 40 before it was launched, they _____ the iPlay 6010 more successfully.
6 Perhaps if Teegan had spent less money on advertising, they _____ the Nitron 40 more competitively.

Videoscripts

Film 1

Part 1 Meet new partners

Michael Hello. My name's Michael Peters, and I'm going to be your guide through the story you're about to see.

For the last ten years, I've been working as a company strategy and management consultant. Companies come in all shapes and sizes, and work in very different ways, depending on their size, corporate culture and national culture. Only one thing is certain – only the fittest survive.

In this story, two very different companies, with different cultures, have come together to work on a major project. They are Paridon and the Bluenet Global Corporation. Paridon is growing every day. It only started six years ago when two 'internet penpals', one in Paris and one in London, realised they could create stunning computer games together. They started a company, and in *that* field, Paridon – the name comes from Paris and *London* – is already a respected brand.

These are the key people in the company. The founders were Jamie Flacks, who has always lived in London, and Philippe Henetier, who was Jamie's internet penpal in Paris. Philippe moved here, to London, two years ago. They are Paridon's Creative Directors. The Managing Director is Hilary Morrison. She's a graduate of the London Business School, where Jamie's father teaches. Jamie and Philippe realised, very early on, that, although they understood computer technology, they needed other skills in order to build a business. Jamie's father introduced them to his star pupil ... and here she is today.

This is a crucial moment in Paridon's history, because this is when the company is branching out into hardware.

Jamie and Philippe have created the music player of tomorrow. It's an almost limitless library of the owner's favourite tracks, as well as a multi-user games machine and a video, text and picture mobile phone, all in one. Its manufacturing cost will be so low that even aggressively competitive pricing will give great margins.

The product doesn't yet have a market name; it's still just 'FH1', which is Paridon's development code for the unit. But 'FH1' could soon become *the* status symbol for everyone in their teens, 20s, 30s and beyond, all over the world.

If plans go right, Paridon can achieve their aims, and they're teaming up with Bluenet to access that global market.

The Bluenet Global Corporation has offices in 26 major cities around the world, and a workforce of nearly 2,000. It's already a leading company in the worldwide marketing of music and video related state-of-the-art designer hardware.

Michael Today, Penelope Bates and Chuck Fenno, CEO of Bluenet, are bringing together, for the first time, the key players from both companies. Everyone knows that first impressions are crucially important.

Hilary Hi, I'm Hilary, Hilary Morrison. You must be Jet. It's good to meet you at last.

Jet Yes, I'm Jet. And it's a pleasure to meet you too, Ms Morrison.

Hilary Oh, 'Hilary', please ...

Jet Thank you. It's a pleasure to meet you too, 'Hilary'.

Chuck Chuck Fenno, CEO of Bluenet.

Philippe Philippe Henetier, one of Paridon's Creative Directors. I'm very pleased to meet you, Mr Fenno.

Chuck 'Chuck', please. And I'm thrilled to meet *you*, Philippe. We wouldn't be here without you today. Nor without you, too, of course, Jamie.

Jamie Thank you, Chuck. But *you're* the most important person here today. You're our key to the global market.

Chuck You're too kind, Jamie. What we have here today is a team – this is a great team.

Penelope Ladies and gentlemen, please take a seat. Thank you, ladies and gentlemen. It's 9:30, so I suggest we get under way, if that's OK. We have a busy day ahead of us.

Firstly, a warm welcome. This is a very important and exciting day for us at Bluenet. We're delighted to be sitting here with Paridon, and especially pleased that our CEO, Chuck, has flown in from New York just to be with us.

Chuck Thank you, Penelope. I'm really glad to be involved. Obviously, as president of our UK division, Penelope will be leading this project. Perhaps you'd like to take us through today's agenda?

Penelope Certainly, Chuck. We have three key items on the agenda today: the marketing strategy, the schedule and, most important of all, the product name. I've planned this meeting, as you can see on the basis of one hour for each of the items. Is everyone happy with this?

Michael The Paridon team members have three main concerns at the moment.

Firstly, they're worried that Bluenet doesn't fully appreciate that this product will appeal to all market segments, not just the teens, 20-somethings and 30-somethings. They're discussing that right now.

Secondly, they're worried about the timescale, on two counts: getting ahead of the competition, and being ready in time for the most important international trade fair in this business. It takes place in just six months' time.

And thirdly, they're anxious about whether they can persuade Bluenet that the name they've chosen for the product is the right one. None of the promotional work can begin until that is agreed.

Philippe Yes, but surely we have to have a separate campaign designed for each market segment, and a specific team of people dealing with each segment, a team that really understands that particular segment?

Chuck Sure. Sure. OK. Let's close this discussion. I believe that we're saying the same thing, but that we're saying it differently. Penelope, could you summarise, please?

Penelope Of course. We *will* target every market segment, Philippe, but it's important that we get one market segment to *lead*, to create the trend. That saves us so much time and money. And it is perfectly possible for one team to create and run campaigns for several market segments. Look, is there any way you can find a little time later today to spend with Jet? He can go through the marketing plans in detail with you.

Philippe OK.

Hilary Excellent. And Jamie, could you join them, do you think? It would be really helpful if you could be there.

Jamie No worries.

Hilary Great.

Chuck Excellent. Thanks, Hilary, thanks guys. OK, I'd like to move on to your concerns about timescales. Hilary?

Hilary Thank you, Chuck. Our two points, as you know, are getting FH1 on the market before Teegan launches its new model, the Nitron 40, and making sure we've established a really powerful campaign well before the international trade fair in Chicago, which is now less than six months' time. Penelope and I have been talking about this for a while.

Penelope Yes, Hilary and I have discussed this a few times, and Jet has drawn up a plan. Jet.

Jet OK. This is our 'overview' workflow plan, which we can discuss now. But, at the end of this meeting, I'll give each of you a copy of the detailed version of our timetable. It contains information about *everything* we intend to do in each market segment, and shows how much time we've allocated to each task. It also shows each task broken down into steps. I'd like to meet to agree those plans by the end of next week, if possible.

Michael The team from Paridon is now less worried. Jet has explained clearly what has been done, and what still needs to be done – so the Paridon team is confident that Bluenet has devised a plan and schedule to meet their two key deadlines: beating Teegan, their main competitor, to the market – and being ready in time for the Chicago trade fair.

But, what about the name? Over to Jamie.

Jamie We all know that we've got a great product. The question is, what are we going to call it? Now I'm sure you have some great ideas, and, in a moment, I'm going to tell you the suggestions that Philippe and I have come up with. But first, I'd like to run through what we think are the most important ingredients that need to be in a name. So I'll start by listing the key points, and then go on to tell you the names that we'd like to propose.

Firstly, let's think about who we're selling to. Although we want to sell in *every* market segment possible, we've based our thinking on the market segment most likely to be interested from the very beginning: the 20-somethings who already have equipment like this. The majority of them are male.

It's important to remember that 'geeky' expressions like 'Vi' for 'virtual reality' and 'i' for internet work very well for this market segment – and so do words like 'Elite', 'Master' and 'Prestige'. Also, these buyers like to think that they're getting the very latest version of anything too. So a new product in this market always starts with a number – never a zero or one, but nearly always with a five or six.

Secondly ...

... OK, let's move on to our suggestions.

We've ended up with a shortlist of three. One, 'the EPLA 5155', which stands for Elite Personal Life Assistant; another 'the EViMM 5050', which stands for Elite Virtual Media Master. They both have the right kind of words in them, and they sort of tell you what the product does. But they're not really snappy.

Number three, we believe, is the best. It's 'the iPlay 6010'. The 'i' says it all: this is the machine of the internet, and it's for *play*, for your leisure time, and just think what you can do with it ...

Jet Fantastic slogans ... 'Work hard, play hard' ... 'i play, u play'...

Penelope It's good.

Jamie So that's our top choice, and that's our proposal. The 'iPlay 6010'. It has everything. It has the key word in it, it really flows as spoken language, it tells you exactly what the product will do for you, and the '6010' makes it, straight away, two generations newer than the Nitron 40 – right from the start – and it's a great name for advertising slogans.

Chuck It's good, very good. Jamie, Philippe, excellent thinking.

Film 2

Part 2 Deal with problems

Hilary That would be fantastic. No, ten o'clock's fine. Thanks. I'll see you then. Bye. OK, let's start! Thanks for coming along, guys; I know how busy you are ...

Jamie No worries.

Philippe It's fine by me, Hilary. It's *very* important. Obviously.

Hilary Great. I think it's really important that we evaluate the meetings we've had with Bluenet ... look at what we've achieved so far – which I think is a lot – and decide how we want to progress.

Jamie OK.

Hilary So, let's start with what we've already achieved. Philippe, what do you think the key outcomes of the first meeting were?

Philippe Well, I'm confident that we've really improved the understanding that Bluenet has of the iPlay, and that's great. But more importantly, I think they now realise just how good our competitors are – both at developing and perfecting their products, and at getting them onto the market quickly.

Hilary Excellent. And Jamie, how was the meeting with Jet?

Jamie Very encouraging. Bluenet certainly have put good people on the project, and they've got every market segment covered. But I'm still worried that there won't be enough people. Jet said that he'd pass that on.

Hilary And he obviously did. Chuck rang me yesterday to talk about that, and he's agreed to increase the numbers of people in the teams working on the iPlay in all the market segments. But he was very clear about the impact on the bottom line. It's going to increase the marketing costs by about 12 per cent which, at the end of the day, is going to decrease profits.

Jamie Well, Jet asked us if we could bring forward the final product tests by a couple of weeks, which means that either Philippe and I work 26 hours a day, or we get some more support staff!

Philippe ... which will increase our costs as well.

Jamie But we agreed to it.

Hilary And you were right to, Jamie; I really don't believe we have any option. This is make or break now. Look, let's sum up and move forward. What we need to do now is focus on building a really good relationship with Bluenet. Their way of working may be different to ours, but they're great at what they do, and we need them. What we must do, is be *absolutely* clear about what we're doing at each stage of the project, so there are no misunderstandings.

Jamie Yes.

Hilary And we must always keep our objectives in mind, and make sure that we review progress regularly. OK?

Jamie Sure.

Philippe OK.

Hilary Talking of progress, how is the iPlay coming along?

Philippe Fine.

Michael Fine? Mmmm ... actually, it's not. There has been real progress in some areas; for example, both companies have agreed to devote the additional resources needed to support the promotion and delivery schedules, and the communication between Paridon and Bluenet is improving all the time.

But innovation means risk, and new technology is no respecter of timetables, as Philippe is about to find out.

Graham Kimberley Hello. Graham Kimberley.

Philippe Good afternoon, Graham. This is Philippe Henetier from Paridon.

Graham Ah, good afternoon, Philippe. Thank you for returning my call so quickly.

Philippe Can I just clarify what you said, that the screens for the units cannot be ready for another 17 or 18 weeks?

Graham Yes, I'm sorry. Well, not in the quantity you wanted. We've encountered a problem with the Silicote-F.

Philippe In that case, we have a big problem – no, we have a very big problem. We have to launch, worldwide, immediately after Chicago.

Graham I know. I'm sorry.

Philippe And there's nothing you can do? Nothing?

Graham No, not in time.

Philippe C'est incroyable. Graham, this means that Teegan will beat us.

Graham Sorry, Philippe. We've done our best.

Philippe Right. You understand, Graham, that we'll have to use another supplier for the screens now ... and that we'll have to enforce the default clauses in our contract with you?

Graham Of course, I understand. I'm just very sorry that this has happened.

Philippe So am I, Graham, but we have to move forward. I'll speak to you soon. Goodbye Graham.

Graham Goodbye, Philippe.

Hilary Yes, Chuck, I know. We have to solve this quickly – and get it right. Not surprisingly, the bank's not very happy. Of course, I'll let you know straight away.

Michael The problem with the screens is a major one. Paridon has a solution, but it's going to increase the demands on their cashflow and borrowing, because Hytek, that's Graham Kimberley's company, had originally won the contract by quoting lower than its competitors. Hilary has managed to convince the bank, in principle at least, to lend Paridon the extra money that it needs, but the bank will need to be convinced that it won't go wrong again. Jamie is going to present for them this afternoon and he's about to do a rehearsal.

Hilary Hi, Jamie. Thank you so much – and you too, Philippe – for pulling this information together so quickly. Well done guys. OK, Jamie, are you ready?

Jamie Yes.

Hilary Don't be nervous. You're great at presentations. And this is only a rehearsal.

Jamie I know, but it's hardly good news I'm breaking, is it?

OK. We all know the problem: Hytek can't get the iPlay screen right in time, in the quantity we need. If we don't solve this problem, we can't launch at Chicago and all the time and money we've spent so far will have been wasted. So, I've checked out two other companies that we could seriously consider ... Crowne and Animex Industries. We know them, of course. And we believe that both Crowne and Animex would now be only between five and ten per cent more than Hytek, not nearly the 30 per cent they were before.

If we look at market share, they command 23 and 26 per cent, so no significant difference there. So we looked at customer satisfaction. Each company included the results of a customer satisfaction survey in its annual report, and, again, they were very similar, although with Crowne slightly in front.

Hilary How do we know they haven't just made up the results?

Jamie Each company's auditors verified them.

Hilary OK, I suppose we have to trust that.

Jamie But these are the most interesting charts of all. We compared eight products that came onto the market last year with important parts supplied by one of these two companies. Then we looked at the schedule, the time it took from the initial idea being announced to the launch.

It looks like this. These are the four products that came onto the market in the shortest time – and all four of them had key parts or processes supplied by Crowne. It looks as if Crowne is the one to turn to when the schedule is really tight.

Hilary Interesting.

Jamie And I phoned a friend who works at one of the companies Crowne supplies …

Hilary And …?

Jamie Well, he said they were excellent – more expensive than the others, but changes of schedule never seem to worry them.

Hilary So your recommendation is Crowne, Jamie?

Jamie Without a doubt. It may cost us more, but I don't see any other solution and I really do think we can have confidence in Crowne delivering.

Hilary OK. You've convinced me. Now let's go and watch you work your magic on the bank!

Film 2

Part 3 Find solutions

Michael The teams have overcome the problem with the screens, and, with an increase of only seven per cent on the cost, much less expensively than they had feared. Jamie's presentation went well, and the bank is willing to service this extra borrowing.

But the teams now have a new crisis. Jet has found out that Teegan, Paridon's main competitor, is launching its own new product, the Nitron 40, which is very similar to the iPlay 6010, with a huge, global advertising campaign, two months earlier than had been planned. That's two months before the Chicago trade fair. Teegan is then going to make its product available in its own retail outlets one month before the fair in Chicago. Now, of course, Paridon doesn't have its own retail outlets, so it can't compete in that way.

Although everyone at Bluenet and Paridon is 100 per cent confident that the iPlay 6010 is the much better product, this strategy by Teegan could be a marketing disaster for them.

Penelope OK. Let's sort out this problem. But first, let's agree what the problem is. Teegan is launching first. Fact. Its Nitron 40 is good, but not as good as the iPlay 6010. Fact. Important fact. So what's going to be the most damaging result of their early launch, bearing in mind that potential buyers won't be able to compare the two products physically?

Hilary Early adopters will want to get hold of a product like this as soon as any version from a respected brand is available. Even if they think a better one is coming?

Penelope That's right.

Chuck So we have to stop them. We have to make them wait.

Philippe And how exactly do we do that?

Penelope Well, *do* we make them wait? What would happen if we launched even earlier than Teegan, and created such excitement that even the impatient 20-somethings would wait?

Jamie Mm, interesting. The iPlay *could* be ready early, now that we've got the new screen supplier in place, but Chicago is the place to launch it. It's the world's marketplace for this sort of thing. It's where we'd get the most impact.

Jet That's true. Everyone in this business goes to Chicago.

Penelope OK. So, if we launch early – with some kind of 'preview' giving the message – all we really have is a communication problem. Nothing more. We just need to work out how to have maximum impact, on a global scale, with minimal cost and very little time. It's a bit of a challenge, I admit.

Chuck More than a bit, I think, Penelope.

Penelope True. But let's think how we meet it. OK, let's have some ideas. Think freely, whatever comes into your heads.

Jamie How about setting up our own trade fair? Before Chicago?

Philippe Impossible!

Penelope Nothing's impossible in a brainstorming session, Philippe!

Jamie OK. Why don't we hold special preview events of our own?

Jet One in every city where we have offices? That would be global.

Chuck Interesting. Yes, I like that!

Jet What if people could actually get their hands on an iPlay when the Nitron is still only on their TV screens, in commercials?

Philippe That's interesting. Go on!

Jet Potential buyers would only see the commercials for the Nitron, but what if they could actually *try* an iPlay? Hold one, test one?

Hilary We'd never have the distribution outlets set up in time.

Jet No, I don't mean have any for sale ... these would be product trials of prototypes, but *public* trials. Very public!

Chuck If we make these events really exclusive, then we'll create a cult straight away.

Jet It's almost a textbook marketing mix. With all its features, the iPlay is way ahead as a *product*; these previews and the impact at Chicago will give us amazing *promotion*; we will, very soon, have global distribution and maximum *availability* and, even maintaining good margins, we can *price* competitively.

Chuck Better features. Great value for money. And we start with the image of being very exclusive. I like it. I really like it!

Michael Tomorrow, Jet is going to make a presentation to the boss of the Bluenet office in every country where the iPlay 6010 is having its early launch. They've all been summoned to London, urgently, but they've no idea what it's about. This is information that's too sensitive to deal with in a conference call or in e-mails. This afternoon, Jet is rehearsing his presentation to both the Bluenet and Paridon teams.

Jet Teegan has beaten us to it. Time to give up!

As if we would ever give up! On anything. Teegan *has* beaten us, but only in time. Let me explain.

What would you do if your client had spent years developing a crucially important new product ... and then their main competitor suddenly shot ahead?

Well, that's what happened to us, to you, only 72 hours ago, when we discovered that Teegan was going to launch its Nitron 40 two months earlier than planned. And it's going to launch it with a *global* campaign of big budget, top quality commercials – the full works!

And Teegan's big advantage, of course, is that it has its own retail outlets, and the Nitron 40 will be available in those outlets in five months' time, one month after the commercials start to air.

We're going make people all round the world desperate to get hold of an iPlay 6010, so excited that they won't even *think* about buying a Nitron 40. And how are we – you – going to do it?

Well, before the Nitron 40 gets into the stores, each of you is going to hold a very special and exclusive launch event. It will be for key industry correspondents and selected consumers in your country. At these events, they will actually be able to use an iPlay 6010, there and then. It won't be available on the market until immediately after Chicago. But these launches will mean that, by then, everyone will know about it. They will know that it is *far* superior to the Nitron, which will look out of date before it even gets onto the market.

So, that's our objective: not to sell the iPlay – not yet – because we can't, but to stop people buying the Nitron.

This is a spoiler strategy, make no mistake. And if we go for it with real aggression, it will work!

Michael Jet has been extremely clear, and has used a structure for his presentation that was designed for the maximum effect. It defined the problem with dramatic impact, but made the solution equally dramatic and, in fact, exciting. Now that Jet has registered those main points, he's moving onto background information that's vitally important.

Jet But before we discuss the details of this plan, let's just consider a few key background facts.

OK, take a look at this graph. You can see that 25 years ago, the principal buyer of new domestic technological devices was aged 40 and now he is aged only 22. Twenty-five years ago, the buyers were 74 per cent male. Now the male still dominates the market, at 58 per cent, but women constitute 42 per cent. So please remember that when you're doing your guest lists!

And this graph shows the main reasons for purchasing. You can see that 25 years ago, 'practicality and long-lasting qualities' were the main factors in 84 per cent of cases. Today, it is more than the reverse: 90 per cent, yes, 90 per cent, put aesthetic appeal as the top requirement – and, for that, read 'status'. In other words, to be somebody, you must own this. So, please, remember that too!

Michael Presenting to people you know can sometimes be even more difficult than presenting to people you don't. But Jet did well there. He observed all the rules of good presentation. And now he, and the rest of the team, can relax. Chuck is taking them all out to dinner.

Chuck So, how did you two actually meet?

Jamie On the internet.

Chuck Really? I thought that was just a story.

Philippe No, it's true. Some people think it's rather sad, actually.

Chuck Well, I don't. I mean, it's led to a great friendship and a terrific business. And it's marvellous to have your hobby as a job.

Jet Have you been to this restaurant before, Hilary?

Hilary No, have you?

Jet Yes. I really like it.

Hilary This reminds me of some of the restaurants in Paris.

Jet You're really lucky having offices there. I've always wanted to visit Paris.

Hilary I tell you what, next time we're going to the Paris office, we'll invite you along.

Jet Oh, great!

Hilary And you can buy us all dinner!

Chuck Oh come on, Penelope. Surely you must have done something other than sport in your teens?

Penelope Nothing worth mentioning, really. Then I went to university, then on to work. And here I am now.

Chuck Oh, go on!

Penelope No, really.

Chuck Well, you're doing a great job.

Hey, listen, guys. It's been great talking to you but I really must be getting back to my hotel. Now's the time I have to start calling the States. I love time zones!

Film 3

Part 4 Reach agreement

Michael All is going well for the pre-Chicago launches of the iPlay 6010, but there's been a disagreement that the teams need to resolve immediately. Hilary telephoned the Bluenet team and asked them to come to a meeting straight away. Chuck couldn't change his plans at such short notice, but Penelope and Jet could, and they've arrived now.

Hilary Thanks so much for coming over.

Penelope No problem. It was obviously important.

Hilary The thing is that we're worried about your plans for previewing the iPlay package.

Penelope Really?

Hilary Yes. We can understand why you want to do it the way you've proposed, but we think that another approach will be much more successful. And we'd like to explain why.

Penelope OK. We think we've got the correct formula, but let's hear what you have to say.

Hilary Great. I'll hand over to Jamie and Philippe, who've been working on this. Jamie....

Jamie Thanks, Hilary. OK, Penelope, Jet, we appreciate that it's an interesting idea to present the iPlay package to the guests at the launch as 'work in progress', and with no fancy documents, no packaging, etc ...

Philippe We understand that this is meant to make them feel that they are being let into a secret, that they really are in at the early stage, that it's a privilege to be there.

Penelope Exactly.

Philippe And it could possibly work.

Jet Come on, Philippe. Much more than 'possibly'.

Philippe Well, we don't think it's such a good idea.

Jamie Look, we understand that there are very good reasons for your strategy, but we really do think that it's too risky an approach. And we'd like to go through the reasons with you.

Penelope Please do.

Philippe As we all know, the key competitive advantages that we have are that the iPlay has many more product features than the Nitron 40.

Jamie The guests at the launches will be able to see and try the iPlay and, so, will experience its features ...

Philippe ... but what they take away with them, the marketing materials, are what they're going to show to other people. And we believe that those materials must be as good, and exciting, and as finished as possible.

Jamie Well designed materials, with superb quality visuals – especially photographs – will make the whole concept of the iPlay look much more attractive ...

Philippe ... and we have to remember that the commercials for the Nitron 40 will be airing by then, and will definitely look good, brilliant, in fact.

Penelope Fair point.

Philippe Photographs and diagrams that show the iPlay in use will make the launch guests think about how they would use it. They will start to think of themselves as owners. And if they see beautifully produced marketing materials …

Jamie … as well as a stunning unit, they'll be even more impressed by the iPlay.

Penelope Hmmm, OK.

Jet But what you're suggesting is going to add a great deal to our costs.

Jamie But the commercial campaign is costing Teegan millions …

Philippe … and we're going to come onto the market for a fraction of the cost. Even if we doubled our budget, we'd still spend less than Teegan is spending on their advertising campaign. And we can have fantastic literature.

Jamie It has to be very, very good. It has to demonstrate all our values. We're trying to start a new brand here.

Jet But we still need to make sure that the guests at the previews feel it's exclusive.

Penelope What you've said, Jamie, Philippe, does make sense, I agree. Thank you. Now, let's not waste any time. Jet, could you get some costings for the higher quality literature, please? And also, I'd like you to think more about the launches. How do we make sure our guests still feel it's secret and exclusive? You always come up with such great ideas!

Michael With that decision made, and fully supported by Chuck, who couldn't attend, the next task is to get these materials produced to the very high quality that Jamie and Philippe have in mind. Of course, the problem with that is that the suppliers know the team is short of time on this.

In fact, they're very short of time, and it may be difficult to get everything done. In circumstances like this, it's vitally important that you negotiate well to avoid having to pay too much to people who know that you possibly have problems.

Even though she can be very tough, Penelope is an excellent negotiator. In this case, what she really wants is the quality that Byrne Graphics can guarantee, but, at the moment, their price is far too high.

Penelope The thing is, Geoff, as I said on the phone, we're all a little surprised at the quote you've given us for printing the re-worked visuals. It's very high.

Geoff I know, Penelope. The problem is I'm going to have to hire in more people to get this done in time, and the materials we're going to have to use are very expensive. I mean, you want the best, don't you?

Penelope Of course we do, Geoff, which is why we always come to you.

Geoff And I want to give you the best, Penelope. But I can't lose money.

Penelope I wouldn't want you to, Geoff. But I've got problems here. Your quote is about 12 per cent more than other companies' and, although I keep saying that the quality you offer is worth paying for, I'm coming under pressure to use someone less expensive.

Geoff I can understand, but …

Penelope And that might cause us problems in the future, if we want to come back to you when we've started building a relationship with another supplier.

Geoff I just don't know how we can cut the costs any more, Penelope. Believe me, I've tried. There's so little time. The other companies have more staff, so they don't need to hire in extra people – and I can't do anything about the cost of the materials, except use cheaper ones, of course, which you wouldn't like.

Penelope Look, Geoff. I want to stay with you, if we can. But we have to benefit too. So here's a suggestion. What if we confirm the other two jobs that you've quoted for recently, the car launch and the holiday company, would that help you? You could get staff in for three jobs instead of one, and get a better price on the stock for a much larger order. The holiday company, after all, is a huge job!

Geoff Yes. That would help. A lot, I think.

Penelope How much is 'a lot', Geoff? ten per cent?

Geoff Penelope! That's outrageous!

Penelope Well, what's the best you could do?

Geoff I don't know. I'll do my very best, OK? Let's just say it will be at least five per cent. I need to do some calculations, see what deals I can get.

Penelope Fine. But this is very urgent, Geoff. We really need to have a decision soon. When can you get back to me?

Geoff Well, it's 11 o'clock now. Mid afternoon?

Penelope Can you get it to me in writing? Fax or e-mail?

Geoff Yes, of course.

Penelope Perfect. Thanks, Geoff. It's a pleasure doing business with you, as always!

Geoff No. Thank you, Penelope. Thanks for your support.

Michael It's the night before the first launch, and the strategy of exclusivity is already paying off. The Paridon offices have already received calls from people who've asked why they haven't been invited. This has raised important questions which the team are discussing now, and details about the launch tomorrow. As time is running out, this meeting is very important and it is essential that everyone has a chance to give their opinions.

Hilary Thanks for coming. I hope you don't mind having this meeting here. It's just that I told all the staff we'd be immediately available if they have any queries.

Chuck That's fine, Hilary.

Hilary And, as I said in my e-mail, I want everyone to have the chance to raise any issue they want. Hopefully you've had time to prepare for this meeting? Penelope, is there anything you would like to raise?

Penelope Well, yes. People have been phoning to ask why they haven't been invited. Quite important people. If we aren't very careful, we could upset a lot of valuable customers and our whole plan will backfire.

Philippe Can I say something?

Hilary Sure.

Philippe I've already spoken to some of them and explained that we have very limited space, but that we're going to send them the literature tomorrow afternoon, and, because we value their support, which I think is very important to stress ...

Jet But don't you think that ...

Philippe Can I just finish?

Jet Yes, OK.

Philippe I'd just like to say that they're very happy with that.

Jamie That's the reaction I've had too.

Penelope Jet, you wanted to say something?

Jet Yes. Perhaps we could get a bigger venue in one or two of our key cities?

Hilary But it might not seem as exclusive in a large venue. People wouldn't feel special any more.

Jamie There is another way to deal with anyone who still feels left out.

Hilary Go on, Jamie.

Jamie Why don't we arrange a private viewing for anyone who didn't get invited – well anyone really important and influential, who isn't happy just to be mailed? We could invite them just to call into their local Bluenet office, if that's OK with you, Penelope?

Penelope That's fine.

Jamie They'd actually feel even more special, that way – and we avoid any negative PR. And we'd also save money on the bigger venue.

Philippe But that could be hundreds of people!

Jamie I don't think so, Philippe. Just think of who we're talking about. It can only be a very few in each city.

Philippe Erm ... I suppose so.

Jamie We cover all angles this way.

Penelope That sounds all right to me. What do you think, Hilary?

Hilary Fine by me, too. All right then. We agree we won't look for a bigger venue. Instead we'll invite any special customers who weren't invited to the launch to go to a Bluenet office – thank you Penelope – for a personal preview. Is everyone happy with that?

Looks like we have work to do!

Film 5

Part 5 Celebrate success

Michael This is the big day. Preparations are taking place for the launch in London and in all the cities around the world where Bluenet has offices. The final team briefing is taking place, just to clear up any loose ends, and make sure that everyone is 100 per cent clear about what is going to happen.

Chuck Good. Excellent! OK, let's get going because we have very little time. We need to be finished by eight sharp, and we've quite a bit to get through. Our objective this morning is to check that everything is organised and that everyone knows exactly what's going to happen. You all have a copy of the agenda, I hope? Good. I hope that I've included every topic that we need to discuss. Can anyone see if I've missed anything? OK. My plan is to bring this meeting to a conclusion at about ten to eight, assign any necessary actions and get us away at eight. Is that OK with everyone? Good.

Jamie ... and I really think that we need to get them out of their seats and circulating as early as possible.

Chuck Thanks, Jamie. Penelope, security arrangements?

Penelope Yes. This is troubling me. I think it may be a mistake to expect the hotel staff to check invitation passes on their own.

Chuck Why?

Penelope Because they won't know if the person arriving is actually the person on the invitation. They don't know the guests. These aren't people that usually come here. And this is a very commercially sensitive event. We really don't want the wrong people here.

Chuck Good point. Hilary?

Hilary Yes, I think Penelope's right. At least one of us should greet the guests as they arrive, and check that they are who they say they are. Perhaps Jamie? Or I could do it, I suppose ...

Philippe Emmm ...

Chuck Thank you, Hilary. But we need to decide exactly who ...

Philippe I'd like to say something here.

Chuck Er, yes, briefly, please, Philippe.

Philippe I think that I should be at the door, with Jamie, because we know different people

Jamie Good point.

Philippe ... and Hilary must surely be in the reception room with time to talk to people individually. This is, after all, a public relations exercise.

Chuck Another good point. Thank you, Philippe. Agreed?

OK, good. It's now eleven minutes to eight, and time to summarise. We are on schedule, but there's no time for any errors. The rehearsal of the presentation will take place ten minutes earlier than planned because the venue staff need time to vacuum the room after the rehearsal, not before, which we weren't expecting. That's at 9:45. After the rehearsal, we'll have a feedback session in here. That will give the event production team time to do a final check. And we'll need the Audio Visual supervisor in here for that meeting. Jet, please make sure he knows.

Jet OK.

Chuck Jamie and Philippe will do the meeting and greeting at the door, with Hilary waiting in the reception lobby. Any questions? Good. Let's go!

Jet Thank you. Now we know that you're all desperate to get your hands on, and try an iPlay 6010 and we're almost finished, so you won't have to wait much longer!

Michael It's gone really, really well, and Jet's about to finish the presentation. This is his kind of thing, and he does these PR presentations very well. Only right at the end did he make clear to the guests that they would actually have a chance to try one of the units. They weren't expecting that, and it caused a great deal of excitement. It was also agreed that Jet wouldn't even mention the Nitron 40 in the main part of his presentation. He would only hint at it in the conclusion. And he's coming to that moment now.

Jet So, to sum up, some try, but no other music player on the market can match the iPlay. It can be an almost limitless library of your favourite tracks and a wireless, multi-user games machine with no limit on the number of players. And, what's more, it's a video, text and internet mobile phone, all in one. No other phone can be a music library and a games machine ... and no other games machine can be a library and a phone. The iPlay is unique.

And no other machine will now have access to the software that leads the computer games market around the world, the Paridon software, invented by these amazing guys!

You've got the best, so don't even think about the rest!

Beware of imitations! You know what we mean, and who we mean!

Thank you.

Thank you for your attention. We won't keep you any longer, but if there are any questions, we'll be delighted to answer them.

Jet Yes?

Jenny Jenny Michaels from Futureworlds. What else have you got planned?

Jet Sorry - are you asking about the iPlay?

Jenny Yes.

Jet Er, Jamie?

Jamie Well, nothing, for the iPlay 6010 ...

Philippe But we know that our competitors will modify their products to catch up with us …

Jet … at which point, be ready for the iPlay 7000!

Jenny And what's that going to offer?

Jet I think that's one for Jamie and Philippe!

Jamie Well, we really can't say at the moment, I'm afraid. There's lots of ideas in development, but it's still very early days.

Jenny But it's not going to be easy to get our retailers and their customers to wait without really good reasons. The Nitron 40 is pretty good, you know.

Philippe But not as good as the iPlay. I think we all know that.

Jenny If we knew what the 7000 was going to offer, or even what it's likely to offer, we'd have a better chance of getting people to buy the 6010.

Jamie I'm afraid any future plans really are confidential at the moment, but … think holograms.

Jenny Really?

Philippe Yes. Really.

Jet Er, off the record, of course!

Robin Why aren't you advertising?

Jet I'm sorry, we didn't quite catch that. Could you repeat the question and your name, please?

Robin Robin Davis, from Cyberstore. Why aren't you advertising?

Jet We will be, but after Chicago, as we always planned. The publicity we get from Chicago is powerful and free, and you all benefit from it too.

Robin Agreed, but when exactly will you be starting your campaign, and what form is it going to take? We can't look uninformed.

Jet Robin, the moment those plans are finalised, we'll be in touch with you all, and give you every detail you need dates, outlets, style, everything.

Michael And so it's over. The iPlay has been launched, and the Paridon and Bluenet teams will see if their strategy of relying on carefully selected word-of-mouth grapevine publicity is more powerful than a multi-million dollar commercials campaign. But, for now, it's time to celebrate!

Hilary Well, whatever happens now, we gave it our best shot. A brilliant job, everyone – especially you guys from Bluenet. Thanks. You've been great.

Chuck What a first class day. Amazing reports from our branches around the world – and then there was Penelope's fantastic deal with Chicago.

Michael This was the icing on the cake. By mid-afternoon, British time, the launches had taken place in 80 per cent of Bluenet's offices, and the industry wires were buzzing around the world. Never one to miss an opportunity, Penelope decided to call Chicago …

Penelope OK, John, so what we've agreed is this. We will change the second phase of our marketing materials to include a full-page advertisement for your fair, but you will supply all the required artwork. In return for that advertisement, you will change our location at the fair to the premier position in Hall A, for no additional cost to us. I would be right in saying that this is the location normally reserved for a company called Teegan, wouldn't I, John? Of course, I knew that! OK, I'll put it all in an e-mail first thing in the morning. Bye, John. Great doing business with you!

Chuck I'd just like to say, fantastic team, fantastic product!

Jamie And can I just say, Chuck, how much Philippe and I appreciate the faith you've shown in us?

Chuck You deserve it. You should be very proud of what you've achieved. And a special thanks to you, Philippe. I know at times you've found us a bit tough, but you've been terrific.

Philippe I don't know what to say! I was just part of a team.

Jet What a fantastic team!

Penelope It's been great working together.

Hilary Well, tomorrow we move on with the 7000. But I think we've deserved a few hours off. Agreed?

All Agreed!

Answer key

Part 1 — Meet new partners

Preview

Introductions
1 c   2 a   3 d   4 b

Video vocabulary
1 status symbol   2 internet penpals   3 crucial
4 key   5 global market   6 star pupil
7 corporate culture   8 agenda

First view
1
1 Paridon and Bluenet
2 Bluenet
3 Jamie's father
4 No, they haven't.
5 New York (in the Bluenet headquarters)
6 FH1, iPlay 6010

2
1 b   2 c   3 a   4 c   5 b   6 c

Second view
1 Jamie Flacks   2 Creative Director
3 Managing Director
1 CEO   2 President of UK division   3 Jet Patel

2
1 ✓   2 ✓   4 ✓   5 ✓   6 ✓

3
1 b   2 e   3 a   4 d   5 c

After viewing
1
1 takes place   2 are sharing   3 meet
4 is preparing   5 are meeting   6 think
7 feels   8 is staying   9 is leaving
10 checks

2
1 a   2 Ø   3 a
4 The   5 the   6 the
7 a   8 Ø   9 a
10 Ø

3
1 are meeting
2 Are we going to go / Are we going
3 are seeing
4 will be
5 will call
6 are you going to present
7 am going to run through
8 are selling
9 am going to tell
10 will be

Part 2 — Deal with problems

Preview

Giving presentations
1, 4 and 5 are good pieces of advice.

Video vocabulary
1 f   2 d   3 i   4 b   5 h   6 e   7 a   8 g   9 c

First view
1
1 F   2 T   3 T   4 F   5 T   6 F   7 F   8 T

2
1 achieved   2 progress   3 fine
4 problem   5 quantity   6 rehearses/practises
7 expensive   8 satisfaction   9 delivery
10 tight

Second view
1
1 e   2 f   3 g   4 a   5 d   6 b   7 c

2
1 Graham Kimberley   2 problem   3 the screens
4 17   5 18   6 call

3
1 b   2 c   3 e   4 a   5 d

4
a 3   b 1   c 2

5
1 g   2 b   3 e   4 d   5 e

After viewing
1
1 have achieved   2 has encountered
3 have put, have got   4 passed, phoned
5 agreed   6 have improved

2
1 b   2 c   3 c   4 b   5 a   6 b

3
1 the lowest   2 the shortest
3 better   4 the best
5 more expensive   6 more reliable

Part 3 Find solutions

Preview

Brainstorming
1 participants
2 problem
3 criticise
4 atmosphere, participation
5 valued
6 sensitive
7 growing, forgotten
8 Shortlist

Video vocabulary

1
1 d 2 a 3 i 4 h 5 g
6 b 7 e 8 c 9 f

2
1 retail outlets 2 Early adopters
3 global scale 4 advertising campaign
5 marketing disaster 6 aesthetic appeal
7 spoiler strategy 8 potential buyers
9 Competitive pricing

First view

1
1 Teegan is launching the Nitron 40 earlier than planned.
2 One month before the Chicago trade fair.
3 Because they would get the most impact if they launched it at the trade fair.
4 Key industry correspondents and selected consumers in every country where Bluenet has offices.
5 At all the Bluenet offices around the world.
6 The number of women buying new technological devices has increased dramatically over the last 25 years and women now represent 42 per cent of the market.
7 Aesthetic appeal.
8 How Jamie and Philippe first met.

2
1 T 2 F 3 F 4 T 5 F 6 T 7 F 8 T

Second view

1
Ideas 1 and 4 are mentioned. Idea 4 is accepted.

2
1 ✓ 2 ✓ 7 ✓

3
Jet does 1, 2, 4 and 7

4
1 ✓ 3 ✓ 4 ✓

5
Students' own answers.

6
Topics 3 (how Philippe and Jamie met), and 7 (the restaurant) are talked about; there is an indication that Chuck and Penelope have talked about 4 (Penelope did some sport in her teens) and Chuck tries to get Penelope to talk about 8 (her teenage years). Penelope shows signs of being uncomfortable about being asked questions about her teenage years.

After viewing

1
1 e 2 c 3 d 4 f 5 a 6 b

2
1 The number of women buying technological devices increased sharply.
2 Sales of the iPlay 6010 fluctuated slightly over a three-month period.
3 Sales dropped slightly in July.
4 Prices rose steadily in 2005.
5 Interest in the Nitron 40 declined gradually after the initial excitement.
6 Sales of the Nitron 40 and the iPlay 6010 were strikingly similar in the first two months.

3
1 which 2 which 3 where ✓
4 where ✓ 5 who ✓ 6 who ✓
7 when 8 who ✓

Part 4 Reach agreement

Preview

Presenting a structured argument
1 c 2 f 3 d 4 b 5 a 6 e

Video vocabulary

1
1 an approach 2 a problem 3 a problem
4 a feature 5 a disagreement 6 a house
7 a strategy 8 a negotiation

2
1, 2 strategy, proposal 3 issue 4 argument
5 argument 6 decision

First view

1
1 P 2 P 3 B 4 P 5 P 6 P 7 B 8 B

2
1 a 2 c 3 b 4 a 5 a 6 a 7 b 8 c

Second view

1
2 ✓ 3 ✓ 4 ✓ 5 ✓

2
1 ✓ 3 ✓ 4 ✓ 5 ✓ 7 ✓
1 c 3 e 4 a 5 d 7 b

3
1 e 2 f 3 d 4 h 5 b
6 g 7 j 8 a 9 c 10 i

After viewing

1
1 f 2 d 3 a 4 e 5 c 6 b

2
1 to resolve	2 previewing	3 to do
4 to make	5 to think	6 to start
7 to get	8 printing	9 saying, paying
10 to use	11 building	12 having

3
1 need	2 have to	3 need to
4 has to	5 ought	6 should
7 must	8 mustn't	9 should
10 doesn't have to		

Part 5 Celebrate success

Preview

Summarising
1 market	2 developed/invented	3 problems
4 supplier	5 two	6 preview
7 customers	8 commercials	

Video vocabulary

1
1 e 2 f 3 b 4 a 5 c 6 d

2
1 a 2 b 3 a 4 c 5 a 6 b 7 c

First view

1
1 F 2 T 3 F 4 T 5 F 6 T 7 T 8 F

2
1 b 2 g 3 f 4 d 5 e 6 c 7 a

Second view

1
The assessment is probably a fair one.

2
1 c 2 b 3 b 4 b 5 d 6 c 7 a 8 c

3
1 a 2 a and b 3 b 4 c

4
1 What else have you got planned? and Why aren't you advertising?
2 He didn't hear the second question about advertising.
3 He didn't understand the first question about future plans.
4 He answered the question about advertising.
5 Jamie (with some help from Philippe) answered the question on future plans.

5
1 g and c 2 e and f 3 a and d 4 b

After viewing

1
Chuck said that we were on schedule, but there was no time for any errors. He explained that the rehearsal of the presentation would take place ten minutes earlier than planned because the venue staff needed time to vacuum the room *after* the rehearsal, not before, which we hadn't been expecting. He said that that would be at 9:45. After the rehearsal, we would have a feedback session. That would give the event production team a chance to do a final check. He also said that we would need the Audio Visual supervisor in there for that meeting. He asked Jet to make sure he knew.

2
1 It was agreed that Jet wouldn't even mention the Nitron 40 in the main part of his presentation.
2 The moment those plans are finalised, we'll be in touch with you all.
3 By mid-afternoon the launches had been held in 80 per cent of Bluenet's offices.
4 The decision to call Chicago and get a better stand at the fair was made by Penelope.
5 Now that the iPlay has been launched, let's go out to celebrate.

3
1 should have designed
2 should have made
3 should have included
4 should have spent
5 might have challenged
6 could have priced

Pearson Education Limited
Edinburgh Gate
Harlow
Essex CM20 2JE
England

And Associated Companies throughout the World.

www.intelligent-business.org

© Pearson Education Limited 2005

The right of Helena Gomm to be identified as author of this Work has been asserted by him/her in accordance with the Copyright, Designs and Patents Act 1988

All rights reserved; no part of this publication may be reproduced, stored in a retrieval system, or transmitted in any form or by any means, electronic, mechanical, photocopying, recording, or otherwise without the prior written permission of the Publishers.

Second impression 2006

ISBN-13:978-0-582-84799-6
ISBN-10:0-582-84799-0

Set in Economist roman 10.5/12.5 pt

Printed by Graficas Estella, Spain

Contents

Introduction	2
General rules	3
Organising your writing	4
Punctuation	6
Numbers	8
Spelling	9
British / US English	10
Abbreviations	12
Text messages	13
Job titles	14
Avoiding errors	15
Letters	16
Emails	20
Faxes	22
Memos	24
Minutes	26
Short reports	28
Press releases	30

Introduction

Learning to write well in a foreign language is one of the most difficult challenges facing the language learner. Finding the right words, using an appropriate style, respecting conventional formats, and correct spelling are just a few of the areas to consider.

Effective writing plays an increasingly important role in today's business world. A clear and concise style ensures that essential information is both understood and acted upon.

The *Intelligent Business Style Guide* is designed to help business students and practising professionals to become more familiar with the styles and techniques of business writing. It contains samples of the most common types of business document, such as letters, emails, memos, faxes, minutes and reports, all with full explanations of usage, language style and layout.

There is also advice on other areas, including abbreviations and text messages, punctuation and how to avoid common mistakes.

We hope that you find the *Style Guide* useful and that it helps you to become more proficient in your business writing.

General rules

Before you write
Ask yourself the following questions:

- **Who** am I writing to? This will help you to determine the tone and degree of formality. When writing to superiors or customers, the tone is more formal than with colleagues. However, business communication should never be too informal and should always be polite and courteous.

- **Why** am I writing? It is a good idea to note down your main message / messages before starting. This will help you to stay focused and also to determine which means of communication is most suitable for your purpose, e.g. fax, letter, email.

- **What** does my correspondent know? Never assume the other person has all the background information necessary to understand your message. Think about what details you need to put into your communication to help them understand and act upon it.

When you write
Whether you are writing a full, formal report or a memo to your colleagues, it is generally agreed that the following points should be respected:

- Be brief, clear and concise. Use simple words and phrases and keep sentences short. In other words, use 'plain English'.
  Don't write:
  If there are any points on which you require explanation or further particulars, we shall be glad to furnish such additional details as may be required by telephone.
  Write:
  If you have any questions, please ring.

- Don't be over-technical or use slang or jargon. Your reader may not understand.
  Don't write:
  I'm on a roll here.
  Write:
  The trip has been successful so far.

- Don't overuse people's first names. It is better to use them only in the opening.
  Don't write:
  Dear John, ... so you see, John, ... I'm sure you understand, John, ... etc.

Before you send

- Always read your business communication before you send it. It is a good idea to read it aloud to see if it is communicative and easy to read.

- Ask yourself: Could the person reading this understand and act upon it from a single reading?

- Check for spelling and grammar mistakes. Do not rely on spellcheck software, as it will not correct mistakes like s in *advise* or *practise* when used as nouns – which use c.

Organising your writing

From plan to paragraph

You should start by thinking about the message that you want to communicate. If you have a clear idea of this, it will help to make your writing coherent.

Make notes of all of your ideas and then organise these into topic areas. Each topic area should consist of one of your main ideas and may also include a number of secondary points that you can develop.

Draw up a plan for each topic area and start to think of the order in which you want to present them. The order will depend on how the topic areas relate to each other and on how you wish to develop your message.

Each topic area and any related secondary points will usually form a paragraph. Moving on to a new paragraph shows your readers that you are focusing on a new topic area.

The paragraphs themselves can be organised in different ways: you may want to present the information chronologically or you may prefer to present points in order of importance. Alternatively, you may wish to balance negative and positive points or compare and contrast different ideas.

First draft

If you have prepared your plans well, the writing process should be relatively straightforward. You can now give your full attention to the language that you are going to use. Prepare a first draft of the complete document and then read through it to see which sections need to be improved or rewritten.

Final draft

Before you present the final draft of your text, you should read it through several times, eliminating repetitions and changing ambiguous phrasing. You can also make long sentences shorter or connect short sentences together. When you are happy with the wording of your final document, you should do a final check for any spelling and punctuation mistakes that you may have overlooked.

Compare the first and final drafts opposite. Notice how in the final draft, sentences and wordings have been changed and linking expressions and paragraphs have been used.

Useful phrases and notations

Linking expressions

Using the right linking expressions will allow you to connect your sentences and your paragraphs together smoothly and will show your reader how your message fits together.

Use linking expressions to

- show contrast:
  However ...   On the other hand ...
  ... whereas ...   ... while ...
  ... although ...

- give additional information:
  Moreover ...   In addition ...
  Furthermore ...

- summarise:
  In conclusion ...   To conclude ...
  To sum up ...

- sequence:
  First ...   Next ...
  Lastly ... / Finally ...

- show cause and effect:
  Consequently ...   As a result ...
  Therefore ...   ... thus ...
  ... because of ...   ... due to ...

- give examples:
  ... for example ...   ... for instance ...

- make references:
  the former ...   the latter ...
  ... the one / the ones ...

Notes

Background: Intro to Environmental Action Plan — reasons for implementation (customers / investors need more info, pressure from regulators & governments)

Methods: Audit of products/processes — Recommendations — Implementation

Performance: Products complete, processes incomplete

First draft

Our Environmental Action Plan was launched two years ago to respond to changes in our industry. More and more customers wanted information about our products' environmental impact. The investment community needed more information and tighter regulation and controls (eco labels) were being introduced in markets. The plan was in three phases: the evaluation of our products and processes, recommendations for changes and implementation. These are now complete for products, where we have improved sustainability by eliminating some production materials. For processes, our goal to introduce guidelines and procedures for the group's activities worldwide has not yet been achieved. But we have introduced standard supplier and logistics management processes and made substantial savings. We are proud of what we have achieved and hope to complete the plan next year. This will enhance our reputation as a business which can meet environmental challenges.

Final draft

Only two years ago we took the important decision to launch our Environmental Action Plan in order to respond to important changes within our industry. Both our customers and the investment community were requesting more information about the environmental impact of our products. Furthermore, governments and regulators in our different markets were imposing 'eco labels' – new standards of environmental conformity.

The plan was in three distinct phases: first, an audit of our products and processes, then recommendations for change and finally implementation in our business operations. On the product side, this is now complete and we have reduced the number of production materials, thus making our products more sustainable. However, for business processes, we have not yet completed the introduction of group-wide policies although we have successfully implemented standard supplier and logistics management processes and made substantial savings.

We are extremely proud of what we have achieved in such a short time and we plan to continue implementation of our Environmental Action Plan, thereby enhancing our reputation as a business that is prepared to meet the environmental challenges of the future.

Punctuation

Using the correct punctuation is an essential part of making your writing clear to your readers. Very often when sentences or texts are difficult to read or do not seem to make sense, it is because they contain errors in punctuation. The following punctuation marks are used in English:

• full stop   .

The full stop, or *period* (US Eng), marks the end of a sentence:

The firms expect Brussels to approve the deal within three weeks.

Full stops are also used after some abbreviations and can be used after numbers which appear in lists:

i.e.   1.   2.

• comma   ,

Commas help the reader to pause at the right point in a sentence and to avoid confusing the meaning within a sentence:

After two months of discussions, the fate of the company has been decided.

Commas are also used when a clause is inserted in the middle of a sentence:

Lagardère, the media-to-missiles group, will not become the biggest publisher.

Commas are particularly important when part of a sentence or word group could be interpreted in different ways:

Most important aid should be lavished on the countries that can use it.
Most important, aid should be lavished on the countries that can use it.

The investors said the fund managers were fools.
The investors, said the fund managers, were fools.

• question mark   ?

Question marks are only used at the end of direct questions. They are not used in indirect questions:

How long will Mr Marchionne be able to survive in his present position?
Many analysts are wondering how the new strategy will work.

• exclamation mark   !

Exclamation marks give extra impact to a sentence and show surprise or shock:

Take it or leave it!
No one was expecting that!

• colon   :

Colons indicate that what follows is an illustration or example of what has been referred to before:

The company is in a strong position financially: its shares are now trading at 4.5 times their original price.

Colons can also be used to introduce lists:

The issues that will be discussed are the following:
* *Education*
* *Trade*
* *Governance*

• semi-colon   ;

Semi-colons mark a pause that is longer than a comma and shorter than a full stop:

Getting accurate results with this method is tricky; two different samples will not produce the same result.

- **apostrophe**   '

An apostrophe shows that something either belongs to a particular person or is closely associated with either a person, group of people or with another thing or things:

Messier's biggest mistake was to have underestimated shareholder discontent.

The apostrophe comes before the possessive s with a singular noun, even when the noun itself ends with an s:

London's traffic problems
my boss's office

The apostrophe comes after the final letter of a plural noun ending with an s:

the employees' complaints

But with irregular plural nouns it is followed by the s:

women's preferences

An apostrophe is also used to show that a letter (or letters) is missing:

We'll agree to your offer if you don't make any further demands.

- **inverted commas**   ' ' OR " "

Inverted commas, or *quotation marks*, are used when citing the exact words that somebody used:

'Just 50m out of 750m Africans have a mobile phone. There is much more room for growth,' says Marten Pieters of Celtel.

- **brackets**   ( ) OR [ ]

Brackets, or *parentheses* (US Eng), are used to present additional information:

Lagardère wants to stop making missiles (his firm owns 15% of the European Aerospace Defence and Space group) and instead concentrate on the company's media interests.

- **dash**   —

Dashes introduce explanations and comments that are connected to what precedes and can, like brackets, show interruptions to the flow of a sentence:

The company shouldn't have agreed to the merger – it wasn't in its best interests. Last weekend in Sydney only half of the properties for auction – the most common method of sale in Australia – were actually sold.

- **hyphen**   -

Hyphens connect two words when they are used as compounds:

state-owned
Asia-Pacific region
debt-equity ratio

Numbers

Numbers, figures and quantities are at the heart of all business operations, and much of the work and time of managers is devoted to calculating, measuring, analysing and presenting numerical data. Profit and performance may be the key indicators of how a business is performing but in the day-to-day running of a business figures have to be entered, references established and records and statistics kept. This continuous flow of numerical information is the lifeblood of the organisation. Knowing how to refer to numbers is therefore an essential skill for the manager. The principal ways of doing this are listed below.

General

Figures are usually written as words when they refer to small quantities. However, for larger amounts they are presented in number form:

Fiat has had five chief executives in two years.

Summer Redstone is the 81-year-old who controls the firm with 71% of its voting stock.

0 is written as **nought** or **zero** (mathematics / temperatures) or as **nil** (scores).

Numbers are not generally used in the plural form except to mean *a lot of* and they are then followed by *of*:

Product development cost four thousand euros.
Product development cost thousands of euros.

When using expressions of quantity as adjectives, use hyphens and use the singular form:

a three-million-dollar contract (a contract worth three million dollars)

Dealing with large numbers

Large numbers are generally presented in numerical form and not as words, unless the number is a 'round' one:

5,361   *five thousand*

If an exact figure is not required, a number is often rounded up or down, and a word like *roughly, approximately, almost, nearly, about, around* may be used with it:

2,464,981 = (around) two and a half million

Abbreviated forms are often used for millions but not generally for billions (except in charts, where the abbreviations **m** and **bn** are systematically used):

The group lost 1.9 billion last year and around 400m the year before.

Fractions and percentages

These can be written in numerical form or as complete words.

Fractions		Percentages	
$\frac{1}{2}$	(a) half, one half	50%	fifty per cent
$\frac{1}{3}$	a third, one third	33%	thirty-three per cent

Decimals

Decimal points and denominators can vary depending on the country. In some countries, the decimal point is represented by a comma and in others by a point. *19.312* would be interpreted as 'nineteen point three one two' in some countries (e.g. the UK, the USA) but in others (e.g. France, Germany) as 'nineteen thousand three hundred and twelve'.

Ratios

These are usually written out in full:

The proposal was adopted by nine votes to two.

Spelling

Using correct spelling is a key consideration in all writing, and misspelling words reflects badly on the author of a document, especially if it is for formal business purposes. Although word processors have built-in spellcheckers which will help you to avoid most basic mistakes, there are also many mistakes that spellcheck software will not identify. Some of the principal difficulties of English spelling are outlined below.

Double consonants

The final consonants of some verbs are doubled when -*ing* or -*ed* is added, but only if the verb has a single short vowel preceding the final consonant and, if it is more than one syllable, the stress is on the final syllable:

stop   *stopped*
begin   *beginning*
but:
develop   *developing*

Not doubling a consonant when necessary is confusing when two different words have similar spellings. *Sitting* is the correct -*ing* form of *sit* but *siting* is the correct form of *site*. (If the consonant is not the last letter of the verb, it is not doubled. But notice that the final 'e' is dropped.)

Double vowels

Some words have two different vowels one after the other. The most common examples are *i/e* and *e/i*. The rule is that *i* comes before *e* when the sound of the word is 'ee', but not when the two vowels come after the consonant *c*:

believe   *receive*

As is often the case with English spelling, there are exceptions to the rule:

seize

Similar sounding words

Some words that have the same pronunciation have quite different spellings and meanings. If you include the wrong word, even if it sounds right, you may in fact be using the wrong spelling:

air	heir	
allowed	aloud	
ate	eight	
board	bored	
cent	sent	
fare	fair	
feat	feet	
genes	jeans	
mail	male	
one	won	
pair	pear	
there	their	they're
wait	weight	

Commonly confused words

It is also easy to confuse words that have only minor differences in spelling. Some of the most common are:

advice	advise
ensure	insure
envelop	envelope
formally	formerly
later	latter
passed	past
perspective	prospective
precedence	precedents
prescribe	proscribe
principal	principle
stationary	stationery
practice	practise

British / US English

Although English is the national language of both the United Kingdom and the United States, there are significant differences in the way that the language is used in both countries. The differences concern not only pronunciation but also grammar, vocabulary and spelling. When writing, it is therefore important to know whether you are addressing an international audience or writing specifically for British or US readers. The sections below provide examples of the main differences between British and US English.

Grammar

The main differences concern spoken rather than written English. In US English, it is more common to use the past simple form of a verb where the present perfect form would be used in British English. This is especially true with the adverbs *just*, *yet* and *already*:

Ms Dewer has just informed me that ... (British)
Ms Dewer just informed me that ... (US)

Prepositions

In British English, prepositions are normally included in situations where in US English they are often dropped:

Judith is returning on Wednesday. (British)
Judith is returning Wednesday. (US)

A different preposition is used in some situations:

The head office is in Wall Street. (British)
The head office is on Wall Street. (US)

I often play golf at the weekend. (British)
I often play golf on the weekend. (US)

Jody is at school today. (British)
Jody is in school today. (US)

Letters

The format for letters is generally the same for both the UK and US. However, there are some differences concerning the opening, where a comma or no punctuation is used in British English (*Dear Mr Harding,*) but a colon is used in US English (*Dear Mr Harding:*).

US closings include *Sincerely (yours)*, *Respectfully*, *Cordially*, *Yours truly*, *Very truly yours*, which are not common in British letters (see **Letters**, page 16).

Dates

Dates can be written with the number before or after the month; before (*14 May*) is more common in British English and after (*May 14*) is more common in the US. When dates are written using only figures, the order is always day/month/year in British English, but month/day/year in the US:

10/06/06   10 June 2006 (British)
               6 October 2006 (US)

Numbers

In British English, *and* is used for numbers in the hundreds, but *and* is not usually included in US English:

669   six hundred and sixty-nine (British)
         six hundred sixty-nine (US)

Quantities

British English tends to use metric units (e.g. *metres*) instead of or as well as imperial units (e.g. *yards*), but in the US it is more common to use imperial units only.

Spelling

The major differences in spelling concern double consonants (less frequent in US English) and word endings. Verbs that end in -ise in British English are usually spelt with -ize in US English. Some nouns that end in -re in British English end in -er in US English. In US English, words often finish in -or whereas in British English they take -our.

British	US
catalogue	catalog
cheque	check
colour	color
dialled	dialed
labelled	labeled
manoeuvre	maneuver
metre	meter
programme	program
sizeable	sizable

Vocabulary

The following list shows where British and US English have a different word for the same thing.

General words

British	US
bill	check
car park	parking lot
clever	smart
diary (appointments)	calendar
flat	apartment
from ... to ...	through
full stop	period
ground floor	first floor
lawyer	attorney
lend	loan
lift	elevator
motorway	highway
neighbourhood	district
ordinary	regular
pavement	sidewalk
petrol	gas / gasoline
post	mail
queue	line
underground	subway
upmarket	upscale
work out	figure out

Business terms

British	US
balance sheet	statement of financial position
banknote	bill
current account	checking account
creditors	payables
debtors	receivables
depreciation	amortization
estate agent	realtor
land and buildings	real estate
ordinary shares	common stock
profit and loss account	income statement
profit	net income
provisions	allowances
savings and loan association	building society
shareholder	stockholder
shares	stocks
stocks	inventories
turnover	revenue
unit trust	mutual fund

The past forms of some verbs are also different.

British	US
fitted	fit
got	gotten

Abbreviations

Abbreviations are often used to refer to names and titles when their full forms are long or complicated. They are also used for technical terms that are often referred to in a particular profession or industry. They are pronounced giving the individual letters that make up the abbreviation and are sometimes preceded by an article: **The BBC**   **The EU**

In formal written texts abbreviations are often printed out in full, but in informal communication by email and over the phone they are normally left in their short forms.

Common abbreviations used in business

Job titles
CEO	Chief Executive Officer
CFO	Chief Financial Officer
CIO	Chief Information Officer
COO	Chief Operating Officer
MP	Member of Parliament

Organisations
EMU	European Monetary Union
EU	European Union
IMF	International Monetary Fund
IRS	Inland Revenue Service
TUC	Trades Union Congress
UN	United Nations
WTO	World Trade Organisation

Countries
UAE	United Arab Emirates
UK	United Kingdom
USA	United States of America

Companies
BA	British Airways
BMW	Bayerische Motoren Werke
IBM	International Business Machines

Business terms
AGM	Annual General Meeting
B2B	Business to Business
B2C	Business to Consumer
CIF	Cost, Insurance, Freight
FOB	Freight on Board
FY	Fiscal Year
GDP	Gross Domestic Product
HR	Human Resources
IPO	Initial Public Offering
M&A	Mergers and Acquisitions
MBA	Master of Business Administration
MBO	Management Buy Out
P&L	Profit and Loss
PLC	Public Limited Company
R&D	Research and Development
ROI	Return on Investment
SWOT	Strengths, Weaknesses, Opportunities, Threats
TQM	Total Quality Management
USP	Unique Selling Proposition
VAT	Value Added Tax

Measurements
ETA	Estimated Time of Arrival
GMT	Greenwich Mean Time
kg	kilogram
kph	kilometres per hour
lb	pound (weight)

Technology
CAD	Computer Assisted Design
CAM	Computer Assisted Manufacturing
FAQ	Frequently Asked Question
HTML	HyperText Markup Language
PDF	Portable Document Format
RAM	Random Access Memory
ROM	Read Only Memory
WWW	World Wide Web

General
AOB	Any Other Business
ASAP	As Soon As Possible
ID	Identity
i.e.	id est (= that is)
PIN	Personal Identification Number

Another term for abbreviation is acronym, generally used when the abbreviation forms a word:
Laser (**L**ight **A**mplification by **S**timulated **E**mission of **R**adiation)
or can be spoken, e.g. *SWOT, FAQ, ROM* and *PIN* from above.

Text messages

The popularity of mobile phones and instant messaging via phone and computer has led to a new form of abbreviated writing, where words are shortened or even sometimes replaced by individual letters, symbols or numbers. This means that fewer keystrokes are required when composing the message, and it also makes it easier for a condensed message to fit on the limited screen space of a mobile phone. Telephone text messages are usually limited in length to 160 characters, but on computer messages they can be of unlimited length. Text messaging is a very informal way of communicating and is never used for official communication.

Here are some simple guidelines for writing text messages in English. There are also several websites where you can write a message that will be automatically transformed into a text message, e.g. www.lingo2word.com.

The most important rules to remember when abbreviating for a text message are the following:

Words that have the same sound as a letter of the alphabet or a number are represented by that letter or number:

U	you	R	are
2	too / to	4	for
8	ate	Y	why

Words that correspond to a graphic symbol on a keyboard are represented by that symbol:

&	and	@	at
#	number	+	plus	-	minus

The percent symbol (%) represents a double 'o':

    l%k   look

Vowels that normally appear in the middle or at the end of words and syllables are removed:

    KNO know   **CHK** check

Upper case letters can represent both vowels and word endings:

    LYK like   **snd*N*** sending

Standardised abbreviations are used for some words and phrases:

    BCZ because   **LOL** Laugh out loud

By using the colon and bracket keys, graphic representations or 'smileys' can be produced to show the emotions of the writer:

    :-) ☺   :-(☹

Sample text message

Are you coming to Pat's presentation at 3 o'clock?

Hope to see you there!

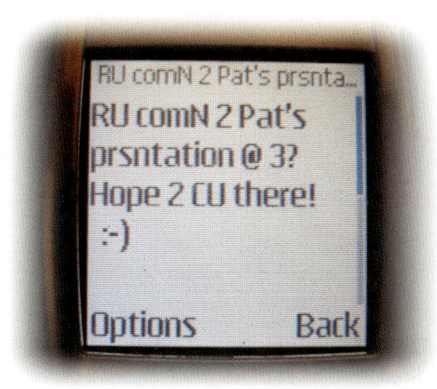

Job titles

In business correspondence, it is important to put a job title after the name of the person you are writing to.

If you do not know the person's name, you can use their title to signal who you wish to contact, addressing them as *sir* or *madam* after that.

Mr Davidson, Personnel Manager

TO: Ms Jennifer Franks, Personal Assistant

The Human Resources Manager
100 Fairview Dock
Liverpool SYX 111

Common job titles

Chairman (of the board) or **President**
the person at the top of the hierarchy

Vice Chairman or **Vice President**
second in the hierarchy

Managing Director or **CEO (Chief Executive Officer)**
in charge of the day-to-day running of the business

Finance Director or **CFO (Chief Financial Officer)**
responsible for all matters concerning finance

Accountant
oversees the book-keeping

Marketing Manager / Director
coordinates all commercial activities

Sales Manager / Director
in charge of the Sales Team

Communications Manager / Director or **CCO (Chief Communications Officer)**
in charge of internal and external communications

Product Manager / Director
manages one of the products in the product portfolio

Legal Affairs Manager / Director
deals with legal matters

IT (Information Technology) Manager / Director or **CIO (Chief Information Officer)**
responsible for all hardware and software in the company

Production Manager / Director
responsible for output

COO (Chief Operating Officer)
responsible for the management of a corporation's day-to-day activities

Plant Manager / Director
in charge of one factory

Supervisor
responsible for a group of workers

Personnel Manager or **Human Resources Manager / Director**
in charge of all matters concerning staff

Research and Development Manager / Director
heads the team that comes up with new ideas and products

Purchasing Manager / Director
deals with suppliers

Facility Manager / Director
in charge of managing and maintaining a company's site (buildings and property)

PA (Personal Assistant)
deals with administrative duties

Avoiding errors

The following is a list of typical mistakes of syntax and grammar that you should look out for when checking what you've written.

Common mistakes	Corrected version
Nouns They need more *sellers*. I am *the responsible of* Marketing. One of the best *product* was …	They need more *salespeople*. I am *responsible for* Marketing. One of the best *products* was …
Pronouns and gender Write to the manager and see if *he* can help.	Write to the manager and see if *he/she / he* or *she* can help.
who / which / whose The manager *which* looks after this … It is the company *who* makes … *Who's* fault was it?	The manager *who* looks after this … It is the company *which* makes … *Whose* fault was it?
Linking words Neither the invoice *or* the order arrived. There is no change *despite* of the new machine.	Neither the invoice *nor* the order arrived. There is no change *in spite of / despite* the new machine.
i.e.* and *e.g. Please cancel our last order, *e.g.* the one for 20 Cartier watches. We stock many brands, *i.e.* Gucci, Dior, etc.	Please cancel our last order, *i.e.* (*that is*) the one for 20 Cartier watches. We stock many brands, *e.g.* (*for example*) Gucci, Dior, etc.
Prepositions I am interested *by* the problem. I've been waiting *the* order *since* two weeks.	I am interested *in* the problem. I've been waiting *for* the order *for* two weeks.
Articles *The* children are most affected by *the* advertising. *A* document you requested has been sent. I am accountant.	Children are most affected by advertising. *The* document you requested has been sent. I am *an* accountant.
Gerunds and infinitives We succeeded *to decide* … We approve *you to do it*. I would like to drive to the conference *instead of to fly*.	We succeeded *in deciding* … We approve *of you doing it*. I would like to drive to the conference *instead of flying*.

Letters

Formal business correspondence is usually done by letter as this leaves a written record which can be kept for reference. Business letters can be of different types with different purposes: to apply for a job, to inform people of developments, to request action, to make an enquiry, to complain, etc. To write a successful business letter you need to use the right tone and to communicate your message to the reader using straightforward language. The way a letter is written reveals a lot about the person who is writing it and it also sends a message about the organisation that he or she is working for. It is, therefore, very important to make sure that the information, layout, style and spelling are all correct before you send it.

Layout

When writing a business letter, you should follow the standard format. The letter opposite shows where the following different components should appear on the page.

a   letterhead / address (but not name) of writer
b   name and address of recipient
c   references
d   date
e   opening
f   subject heading
g   body of the letter
h   closing
i   signature
j   name and job title
k   enclosures

Language styles

Business letters are usually quite formal in style. A conversational style is therefore not appropriate and you should avoid contractions, for example. Try to use verbs in the active and not in the passive form as this will make your letter more dynamic. You should also avoid writing sentences that are too long and that include complicated or unnecessary language. A straightforward letter will get your message across more effectively than a long wordy one. There are certain conventions concerning the correct way to address people and to close your letter.

Opening Letters always start with *Dear ...* followed by the correct form of address. If the letter is going to someone whose name you do not know, it starts with *Dear Sir*, or *Dear Madam*, or *Dear Sir or Madam*, ... But if you do know the name, then you can begin with *Dear Mr / Ms Taylor*, or *Dear Greg Taylor*, ...

Closing Letters are usually closed in standard ways. At the end of your letter you should include a short sentence like *I look forward to hearing from you.* or *Please do not hesitate to contact me if you need further information.* Below that, you should put a closing phrase:

Yours sincerely, (formal, for letters beginning *Dear* + name)

Yours faithfully, (formal, for letters beginning *Dear Sir / Madam*)

Yours truly, / Best regards, / Best wishes, (less formal)

Useful phrases and notations

Following our recent telephone conversation ...
I am writing to you to confirm ...
Thank you for taking the time to ...
Regarding the question of ...
I think you will agree that ...
We are sorry for any inconvenience caused.
I am enclosing full contact details.
enc (shows that something is enclosed)
cc (copy sent to another person)
PS (for additional sentence(s) included after the signature)

Sample letter (of confirmation)

a

HPSempra *Systems* Inc
Conway Industrial Estate
Hartlepool
HPL 7GN
United Kingdom
☏ (+44) 00723 91847
mark@sales/sempra.org

b

Jeanne Eckhart
Optecol
31 Rue Jules Welter
Sanem
L-4902
Luxembourg

c Our ref: TS50.001
d 21 May 200_

e Dear Ms Eckhart,
f **TS50 Delivery**

g

Following our recent telephone conversation, I am writing to you to confirm the arrangements for the delivery of the TS50 test simulator that you ordered through our Brussels office. As I mentioned, a team of engineers from our UK and German offices will be attending the engineering trade show in Hanover from 1–3 July. Our stand at the show will feature a number of new products and test stations, including the first production model of the TS50, which is being shipped over from the UK especially for the event.

The show closes its doors on the evening of the third, and we will arrange for the simulator to be dismantled and packed on to a flatbed truck for delivery to you directly. The transportation will be handled by the same freight company that is looking after the logistics for the show, TransMax, and I am enclosing full contact details with this letter. I have already spoken with their director, Mr Wolfgang Hartung, and he is awaiting confirmation from your side before proceeding with the delivery.

Jane Bradley, the sales engineer who will be in charge of our stand, will be available to travel to your plant on the Thursday after the show, and she will provide assistance with the final installation and configuration of the machine during the following two days.

I will not personally be present at the fair but Ms Bradley will be able to answer any queries you may have. I know that this is an unusual delivery procedure but I think it is an interesting opportunity for you to save the cost and time delay of normal delivery from the UK.

I look forward to doing business with you in the future and thank you again for placing one of the first orders for the TS50. I am sure it will give you complete satisfaction.

h Yours sincerely,

i *Mark Rathbone*

j Mark Rathbone
   Sales Manager

k enc

VAT Registration No. 49238756930-609   Registered offices: HPSempra House, Ludgate Drive, London W11 44SP

Sample letters (of enquiry and reply)

Hoosley Brothers Ltd
42 The Quayside
Dublin 11
Tel 0035378992211
Fax 0035378993311

Hoosleys@office.com

The Export Sales Manager
WalesDepotPlus
2 Docklands Row
Cardiff
RC 8 JK

September 20th 200_

Dear Sir or Madam

We have been given your name by the Chamber of Commerce, as you are an exporter of high quality office supplies. Hoosley Brothers Ltd is a relatively new company dedicated to supplying top of the range stationery and office equipment to businesses throughout Ireland. We have been trading for one year and, having successfully built up an impressive customer base, we are now ready to expand.

We are therefore looking for new, reliable sources of supplies. Any information you can send us on your products, e.g. photos, catalogues, etc., would be greatly appreciated. We would also be grateful if you could provide us with your prices in euros and send us details of your terms and conditions.

Yours faithfully

John Hoosley

John Hoosley
Purchasing Manager

Dear Mr Hoosley,

Thank you for your enquiry of September 20th asking for information about our products. We are very happy to enclose our latest brochure. I would like to draw your attention to our special rates and easy terms of payments for all our new customers this year. I would also like to inform you that it is possible to make purchases online from www.WalesDepotPlus.co.uk.

On behalf of all my team, I'd like to say that we are really looking forward to doing business with Hoosley Brothers in what we hope will be the near future.

If you need any further information, please do not hesitate to contact me.

Yours sincerely,

Sally Thornton

Sally Thornton
Export Sales Manager
s.thornton@WalesDepotPlus.co.uk
enc

Sample letters (of complaint and apology)

Dear Ms Lopez

As someone who has worked with your company for over five years, I was very disappointed when I saw the fliers you produced for our latest campaign.

As our written contract stipulated, we expected full colour photos. Instead, we found that black and white photos had been included in the printed leaflets. I think you will agree that this is an unsatisfactory situation.

We would like you to produce the fliers according to our specifications or provide us with a full refund as quickly as possible.

I look forward to a satisfactory reply.

Yours sincerely

Donna Keller

Donna Keller
Managing Director

Dear Ms Keller

I was very sorry to read the letter sent yesterday dealing with the issue of unsatisfactory printed publicity material. As someone who values your business, I have already begun to rectify the problem.

We will produce full colour leaflets for the end of the week and hope that you will accept our apologies. We will also deduct 20% for any inconvenience caused.

Thank you for your patience.

Yours sincerely

Rosetta Lopez

Rosetta Lopez
Director of Customer Relations

Emails

Email is one of the most commonly-used forms of communication in the international business world. It is used extensively within companies to circulate information, requests, results, instructions, recommendations, minutes of meetings, etc. Email is an effective, rapid and relatively cheap means of communicating with customers and suppliers, both nationally and internationally. Because of the brevity, rapidity and relative informality of emails, it is important to check that all information has been given and that the tone is appropriate.

Layout

The emails opposite show where the following different components appear (though a–g may be in a different order depending on the system being used).

- a name of the person sending the email
- b name of the person / people the email is addressed to
- c person / people who will also receive the mail though it is not addressed directly to them
- d person / people who will receive a copy without other people knowing
- e date (and time)
- f information about the content of the email
- g files, documents, etc. sent separately rather than included in the actual message or body of the email
- h opening
- i body of the email
- j closing
- k name and job title

Language styles

Emails are usually shorter than other forms of communication and the language is simple and concise.

The tone for emails to superiors or people outside the company should be formal. But 'in-house' emails between colleagues can be semi-formal.

When replying to mails, it is important not to reply simply 'yes' or 'no' to questions without referring back to the question and not to use pronouns out of context. However, if returning the sender's email with your reply, by using 'Reply', the sender will have their original to refer to.

Useful phrases and notations

Opening and closing

For semi-formal emails, *Hello* and *Hi* are common openings and *Best wishes* or *All the best* are often used to close.

When writing to several people, you can address the group, e.g. *Dear Project Managers ....*

Formal emails, like letters, start with *Dear Sir / Madam* or *Dear Mr / Ms X* and close with *Yours sincerely* or *Yours faithfully* as appropriate (see **Letters**, page 16).

Exchanging information

Could you mail me the sales figures for last month by 4pm?

Here is the brochure on the new product you asked for (see attachment).

I'm sending you the board's recommendations with this mail.

Please note that today's appraisal meetings have been cancelled. They will be rescheduled for the same times next Monday.

Just to let you know, I sent the attached minutes of Tuesday's Production meeting to all the participants.

Replying to emails

Thank you for your mail.

I got your mail, thanks.

Sorry I didn't get back to you yesterday but ...

Re your request for last month's sales figures ...

I couldn't open the attachment you sent – please resend it as soon as you can ...

Thanks for the information about the appraisal meetings. I'll make a note of it.

Sample emails

Formal

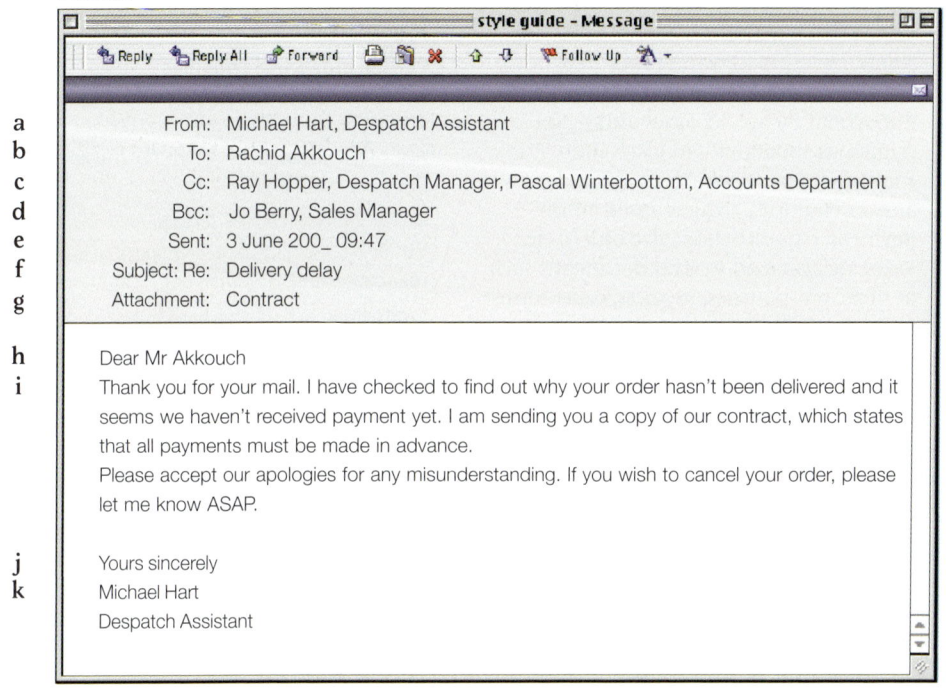

a	From: Michael Hart, Despatch Assistant
b	To: Rachid Akkouch
c	Cc: Ray Hopper, Despatch Manager, Pascal Winterbottom, Accounts Department
d	Bcc: Jo Berry, Sales Manager
e	Sent: 3 June 200_ 09:47
f	Subject: Re: Delivery delay
g	Attachment: Contract

h Dear Mr Akkouch
i Thank you for your mail. I have checked to find out why your order hasn't been delivered and it seems we haven't received payment yet. I am sending you a copy of our contract, which states that all payments must be made in advance.
Please accept our apologies for any misunderstanding. If you wish to cancel your order, please let me know ASAP.

j Yours sincerely
k Michael Hart
Despatch Assistant

Semi-formal

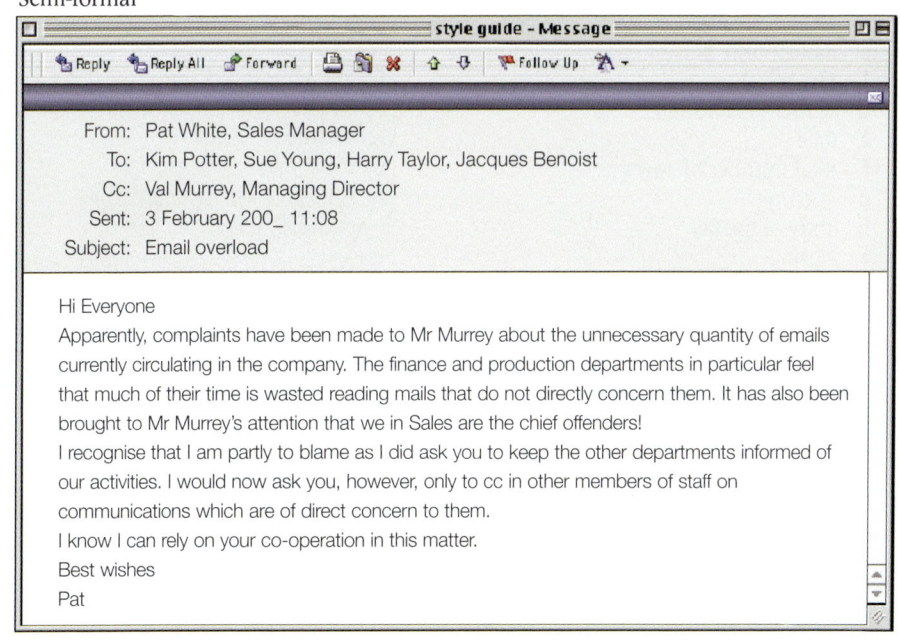

From: Pat White, Sales Manager
To: Kim Potter, Sue Young, Harry Taylor, Jacques Benoist
Cc: Val Murrey, Managing Director
Sent: 3 February 200_ 11:08
Subject: Email overload

Hi Everyone
Apparently, complaints have been made to Mr Murrey about the unnecessary quantity of emails currently circulating in the company. The finance and production departments in particular feel that much of their time is wasted reading mails that do not directly concern them. It has also been brought to Mr Murrey's attention that we in Sales are the chief offenders!
I recognise that I am partly to blame as I did ask you to keep the other departments informed of our activities. I would now ask you, however, only to cc in other members of staff on communications which are of direct concern to them.
I know I can rely on your co-operation in this matter.
Best wishes
Pat

Faxes

Despite the widespread use of emails today, the fax still remains a preferred means of communication for various business functions in a large number of cultures throughout the world. Faxes are sent in numerous situations – to place and confirm orders / bookings, make and answer enquiries, request and confirm payment, request action, give orders, etc. Faxes are also used to send documents such as fliers, prospectuses, invoices, order forms and other business documents. In many cases, documents are faxed first to save time, and the original and / or legal document is sent by post later.

Layout

When writing faxes, you should follow the standard format. The fax opposite shows where the following different components should appear on the page (though b–f may be in a different order).

a   letterhead
b   name of the person / people who will receive the fax
c   name of the person / people sending the fax
d   sender's fax number
e   date
f   total number of pages
g   opening
h   body of the fax
i   closing
j   signature

Language styles

Faxes can be formal or semi-formal depending on their context and form. They can take the form of business letters or emails, or be in note form like memos. The style of language used should be adapted to each situation (see **Letters**, page 16, **Emails**, page 20, **Memos**, page 24).

Useful phrases

Placing orders and bookings
We wish to place an order for / to book …
Thank you for sending the order for …
Please let me know by return fax if you can meet our order / accept our booking.
Kindly indicate earliest possible delivery date.
We look forward to receiving your acknowledgement / shipment / confirmation.

Confirming orders and bookings
Thank you for your order no. 000 / booking dated 5th March …
Your order / booking is receiving our immediate attention.
We trust the quality of our products will meet with your requirements …
You will receive our invoice within the next few days.

Making requests and enquiries
We would be interested in learning more about your products.
Would you mind faxing us your latest price list?
Could you send us, by fax, your conditions and terms of payment.

Giving information and answering enquires
Following / Further to our telephone conversation / meeting …
We thank you for your enquiry.
In accordance with your request for information …
We are happy to fax you the information you require.
Please find herewith the required price list.
After much reflection and discussion, we have decided to do the following …
I trust you will find everything in order.

Requesting action
Cancel …
Send …
Make arrangements for …

Sample fax

a

First Date
Teenage Fashion Designers
Savile Row
London W1 PT 4AZ

Tel: 0020 766776776

Fax Transmission

b **TO:** Ms L Atlan (330144556600)

c **FROM:** Sandy Bradshaw

d **FAX NUMBER:** 0020 7667776767

e **DATE:** Tuesday 18 March

f **NUMBER OF PAGES:** (including this) 12

---

g Dear Ms Atlan,

h Following our meeting this morning, I am faxing you copies of all the documents you require for our loan application.

- Project feasibility study
- Risk evaluation study
- Costs forecast
- Sales forecast
- Complete project budget
- Current balance sheet
- Completed loan application form

I trust that you will find that everything is in order and that you can proceed by filing our application with the bank's board, under the terms we agreed.

I am sending exact specifications and photographs of the new designs separately by courier. These are highly confidential documents and therefore we rely on your utmost discretion.

I would like to take this opportunity to thank you for your continued support and valuable advice on strategy and capital management.

Please let me know by return fax if you need any other papers.

i Yours sincerely,

j *Sandy Bradshaw*

Memos

The word *memorandum*, or *memo* for short, originally meant a reminder or confirmation. Now it has become a very common form of business communication used for a wide variety of messages exchanged between people working in the same organisation. A memo usually focuses on only one specific topic, as in the following examples:

Conveying information Reporting back the minutes of meetings or summaries of brainstorming sessions

Requesting information Asking employees to send in requests for the use of office parking spaces

Giving instructions Telling employees to display identity badges when entering the building

Recommending options Informing people in the company of decisions reached on the best way to solve a company dilemma and recommending that these options be implemented

Layout

When writing memos, you should follow the standard format. The memo opposite shows where the following components should appear on the page.

a   date
b   name of the person / people the memo is addressed to
c   name of the person / people sending the memo
d   information about the content of the memo
e   introduction to the subject matter
f   main points
g   conclusion, often recommending the action to be taken
h   closing, which can be the name or initials of the person sending the memo

Language styles

Although styles vary across cultures and organisations, there are basic rules for memo writing.

The opening is more direct and less formal than in a letter or email, with no greeting such as *Dear ...* and memos usually start with the introduction to the main points. The closing is generally just the initials of the sender.

Memos are less formal than business letters so the tone is neutral and the language simple.

Sentences are usually short and clear, but not brisk and bossy.

Memos often conclude with a request for action.

Useful phrases

Giving information
You will be happy to hear ...
I am / We are delighted to inform you ...
I am delighted to be able to announce ...
I would like to remind you that ...
I have recently been informed ...

Requesting information
I would like to have ...
Could you provide me with ...?
If you have any questions, please ...

Giving instructions
Please read ...
We / I kindly request ...
Make sure that ...
... is permitted only ...

Recommending options
It is recommended that ...
It is in the best interests of ...
Having considered all the alternatives, I / we suggest ...

Sample memo

IElectroTechnika Inc.

MEMO

DATE	8th June
TO	All staff
FROM	Alison Stanford, CIO
SUBJECT	Use of information resources

a
b
c
d

e I have recently been informed by the HR Department of various irregularities in the use of information resources by members of staff in some departments. I have therefore prepared the following memorandum in order to outline company policy regarding the proper use of information resources by all personnel. Please read these guidelines carefully and make sure that you apply them at all times when using information resources.

f 'Information resources' refers to all computers, peripherals and software and electronic mail services that are used as operational components for conducting the company's business.

Permissible use
Although information resources are designed to be used for *official* purposes, it is in the interests of all members of staff to have access to these resources for limited *personal* use. However, all *inappropriate* use will lead to disciplinary proceedings.

Official use
This includes the use of equipment for all activities relating to company business and covers all work-related activities that have been authorised by head office.

Non-official use
Personal use of IT equipment is permitted only on condition that it does not interfere either with a staff member's productivity or affect the productivity of any other member of staff. Short email messages may be sent to colleagues, and internet services may be consulted for limited periods of time in order to obtain news and information.

Inappropriate use
All staff are reminded that they are not permitted to:
- download software to their computers
- use company IT equipment to conduct business for personal gain
- disable virus protection software
- post company information to persons outside the organisation
- make copies of any software installed on computers

g I would also like to remind all staff that their use of information resources may be monitored. It is therefore recommended that you do not use the company's information resources to communicate information that you wish to keep private.

If you have any questions relating to these policy guidelines, you can consult the newly updated FAQ section on our intranet or contact Heidi Wassermann in Human Resources.

h Alison Stanford

Minutes

At every business meeting someone is assigned to 'take the minutes'. This person notes down all the important points made at the meeting and later writes up a clear summary of what was said and decided. It is generally agreed that the minutes should be sent within 24 hours to all the participants and anyone else affected by the content. This ensures that people stay focused on the issues raised and keep future action points clear in their minds. Minutes are sent to make sure that things discussed at meetings actually get done.

Layout

When writing minutes, you should follow the standard format. The minutes opposite show where the following different components should appear on the page.

a subject and date of the meeting
b list of participants
c summary of the chairperson's introduction
d summary of opinions and suggestions exchanged
e action points decided upon, people assigned to each action and deadlines given
f date and time of next meeting

Language styles

The style of language is quite formal. Sentences should be short, clear, concise and easy to read. It is important to summarise only the most important points, not include everything that was said.

Long speeches made at meetings should be summed up, using bullet points for clarity. The minutes of even a long meeting shouldn't be longer than one page.

Useful phrases

Giving the list of participants
Use *Present:* followed by the list.

Summarising the chairperson's introduction
Mr / Ms X opened the meeting with the following points: (+ bullet points)

Summarising a discussion
Mr X was the first to speak / react / contribute and put forward the following ideas: …
Mr Y agreed / expressed approval and added: …
Ms Z disagreed and argued: …
Mr X expressed disapproval / concern and insisted: …
Ms Z accepted that … but defended her position on … However, she conceded that …
Mr X agreed / offered / promised / refused / wanted … (followed by *to* + infinitive)
Mr Y denied / admitted / suggested / recommended … (followed by the gerund)
Ms Z advised / asked / instructed / reminded … (followed by object + *to* + infinitive)
Everyone agreed on …

Action points and deadlines
The following action points were decided upon: …
Mr X will look into / research / draw up a list of / calculate / study … by the end of next week.
Mr Y will speak to … before the 15th.
Ms Z will come to the next meeting with …
Mr X will get back to / send a report to the Finance Committee within the next few days.

Sample minutes

a

Minutes of the Training Budget Meeting
17 November

b

Present: Ms Graham, Managing Director (chairperson); Mr Bhupathi, IT Manager; Ms McKenzie, Human Resources Manager; Mr Wenzel, Finance Director

c

Ms Graham opened the meeting by welcoming all the members and congratulating Ms McKenzie on her first month's performance and on how quickly she'd adapted to our corporate culture. She expressed regret that Ms Amritraj (Export Manager) couldn't attend as she had been called away unexpectedly to deal with a problem with our agent in India.

She then announced that:
- the training budget would be cut by 20% this year
- the HR Department is developing a new cost-cutting strategy and explained that:
a the company would no longer be using the services of Target Training Consultants
b employee development would be more closely linked to business-growth strategies
c all future training would be based on a detailed in-house needs analysis assessment
d training would be strictly limited to staff whose performances are crucial to the company's success.

d

Mr Wenzel was the first to speak and suggested hiring a small independent consultant, whose fees would be considerably lower than those of large firms like Target Training.

Ms McKenzie said that while independent consultants could be cheaper, they would almost certainly fail to ensure that resources were used as effectively as possible. She argued that she would prefer to process all requests and proposals for training personally. She added that any request agreed upon by her predecessor was also currently under review.

Mr Bhupathi expressed concern about his application for training in Java and Java script for his staff. He requested that his application be given priority status. He said that he had already contacted the training centre recommended by Target Training on the advice of our former HR manager.

Ms McKenzie conceded that this training was essential but advised him to resubmit his request asking for cheaper online training.

Mr Bhupathi expressed disapproval of this idea and continued to make a case for using the training centre but he finally agreed to look at what is available online.

e

The following action points were decided upon:
- Mr Bhupathi will draw up a list of e-learning courses currently available and submit a new proposal to the HR Department as soon as possible.
- Ms Graham will develop a detailed needs analysis assessment questionnaire to be distributed to all departments.
- Mr Wenzel will prepare a cost breakdown of the previous year's spending on training and arrange a meeting with Ms McKenzie's department next week to discuss the new cost-cutting measures in more detail.

f

Date of next meeting – 12 December, in the main boardroom

Short reports

Short reports are used to summarise information that has to be communicated to people inside or outside an organisation. They are designed to provide an overview which can be read and assimilated quickly. Many different subjects can be presented in a short report and some of the most common types of short reports are project / progress reports, business proposals and summaries of research or results. Although the length of a short report will vary depending on the amount of information and commentary that it contains, most short reports will be between one and six pages long. They should be clearly structured so that the reader can find the relevant information quickly. Short reports may also include graphic material and are often used as the basis for an oral presentation.

Layout

Title page – indicates the subject that is being dealt with, in large font, with the name and position of the author of the report clearly indicated at the bottom of the page, together with the date of its publication.

Summary – gives a concise presentation of the report, the reasons for writing it, the most important information it contains and a general idea of its main findings. For a short report this can be simply one or two sentences.

Introduction – presents the overview, showing why the report was written and how it has been constructed.

Development section – includes the main body of information which may be divided into several subsections.

Conclusion – presents the results of the report. This might take the form of a recommendation for future action or draw the reader's attention to problems that need to be addressed.

Language styles

Short reports are documents that use a formal writing style. They should not contain contracted verb forms like *it'll* or *don't* but use the full forms of verbs.

The language of the report should be as clear as possible. It is not necessary to use long and complicated sentences or obscure vocabulary. Using simple language in short sentences will make it easier for the reader to assimilate your message.

Try to avoid using the personal pronoun *I* too much. Although it is true that 'you' wrote the report, it will only put your readers off if you refer too often to your own role. Use neutral phrasing instead.

Useful phrases

Connect sentences by

- showing contrast:

  However, this does not necessarily mean that ...

  On the other hand, it is true that ...

  While these results may appear to ...

  Although it is not completely clear from our research ...

- showing cause and effect:

  This has been mainly **due to** ...

  Consequently, additional data was obtained which showed that ...

  As a result of this, it has not been possible to ...

- following on to the next point that you wish to make:

  Moreover, in this particular case ...

  In addition to this, the survey included ...

  Furthermore, we would advise that this would provide an opportunity to ...

Sample report

Acquisition evaluation of PromoVista S.A.

Summary
Michael Paterson Associates has requested an evaluation of the French company PromoVista S.A. with a view to analysing its value as a potential takeover target in order to enhance the international marketing reach of the company's current European operations. The following short report situates the target company in its competitive environment and gives a brief analysis of PromoVista's current position within its industry sector, its strategic positioning and its recent performance, and provides a forecast for the company's future in the short term.

Introduction
Adwise Incorporated looked at the following:
- History
- Competitive environment
- Operational specifics

History
PromoVista was created in the 1930s as a regional point of sale and outdoor advertising provider and has undergone a series of transformations during its eighty-year history. From a small regional operator it has developed into an operation of national and international scope, with subsidiaries in Belgium, Luxembourg and Switzerland. The company currently employs ...

Competitive environment
PromoVista was a monopoly provider of outdoor advertising space until 1985 when BXG Advertising created Proximos, a new outdoor advertising operation in France, to support its strategy of developing national weekly and monthly publications. Proximos is now PromoVista's principal competitor. During the last five years, PromoVista's revenue has continued to increase despite greater competitive pressure, in particular from urban transit space providers and also urban furniture advertisers. However, during this period ...

Operational specifics
PromoVista, unlike its main competitor in its home market, does not rely on outsourcing poster installation and display maintenance work but has its own national teams of specialists who are directly responsible for changing posters and maintaining display sites. In terms of its implantation, the company currently has 120,000 advertising sites positioned essentially in pedestrian shopping zones and town centres. Its research department ...

Conclusion
Analysis of the recent performance of PromoVista shows that the company has performed well over the last five-year period, with profit margins at 20%. We would advise that PromoVista would provide a very interesting growth opportunity. Its position in the market would enable Michael Paterson Associates to enlarge its market share, allowing it to offer enhanced and targeted marketing to advertisers in the markets mentioned above.

Jeremy Hickson
Senior Consultant
Adwise Incorporated

29th August 200_

Press releases

Press releases or news releases are documents that are prepared by the Public Relations departments of corporations and other organisations in order to communicate information to the media. They are designed to arouse media interest by presenting interesting and original information that can be used in an article or a radio or television report. Ideally, the press release can be included in a publication or report with only a minimum of editing. Most press releases are of two basic types: those that inform people about changes (new products, company results, forthcoming events, etc.) and those that try to influence people to adopt a particular point of view (reports, speeches, negotiations, etc.). Press releases are rarely longer than two pages and they follow a standard format.

Format

The basic format for a press release is similar to that of a short article. Start the release with the most important information, which should provide the answers to the questions *who? what? when? where? how?*

In the following paragraphs you can add in more information about the story and include quotations from some of the people involved.

Layout

a letterhead of the organisation that is sending the release
b headline or title (in the present tense)
c an indication of when the information in the release can be published: 'For immediate release' if the news can be published now or 'Embargoed until ...' if it can only be published after a certain date
d date when the release was sent
e main text of the release, divided into paragraphs with paragraph headings where necessary
f ### to show the end of the text
g contact details of the person who wrote the release and information about who to contact for further information

Sample press release

a

ESSprinter   **FasterTracks**

Press Release

b **EsSprinter and FasterTracks Communications announce merger**

c Embargoed until Dec 10

EsSprinter and FasterTracks announced their agreement to merge with a commitment to create a global communications company, offering a comprehensive range of services to consumer business and government customers.

d New York City Dec 5 200_

e EsSprinter and FasterTracks will merge January 15 200_. "The companies are in the enviable position of possessing two incredibly valuable brands," said Daniel Schweitzer, designated Chief Marketing Officer for the new company and FasterTracks' current Vice President, Marketing.

The new strategy and logo will integrate the most valuable assets of each company's identity. Given its broad appeal on the market and its history of innovation, the EsSprinter name has been chosen as the lead name for the new company. The combined company will employ FasterTracks as a product brand within the EsSprinter service portfolio.

The new logo will blend elements of EsSprinter's bold red and yellow colours with the strong visual from FasterTracks' famous trademark, creating a powerful symbol for the new dynamic brand.

The inauguration ceremony, which takes place at the EsSprinter headquarters in New York on Jan 20 200_ will be attended by over 3,000 guests from all over the world.

f ###

g For more information about the inauguration and to obtain high resolution photos of the new logo, please contact:

John Pemberton
T: (+01) 212 555-4500
F: (+01) 212 555-4599
Johnpemberton@essprinterfast.com

Pearson Education Limited
Edinburgh Gate
Harlow
Essex CM20 2JE
England
and Associated Companies throughout the world.

www.longman.com

© Pearson Education Limited 2006

The right of Tonya Trappe and Graham Tullis to be identified as authors of this Work has been asserted by them in accordance with the Copyright, Designs and Patents Act 1988.

All rights reserved; no part of this publication may be reproduced, stored in a retrieval system, or transmitted in any form or by any means, electronic, mechanical, photocopying, recording, or otherwise without the prior written permission of the Publishers.

First published 2006

Available only as part of Intelligent Business Upper Intermediate Coursebook
Not for separate sale.

Set in Economist Roman 10.5 /12.5

Printed in Slovakia by Neografia

Designed by Wooden Ark